Spacecraft interact with the space environment in ways that may affect
the operation of the spacecraft as well as any scientific experiments that
are carried out from the spacecraft platform. In turn, the study of these
interactions provides information on the space environment. Adverse
environmental effects, such as the effect of the radiation belts on
electronics and spacecraft charging from the magnetospheric plasma,
mean that designers need to understand interactive phenomena to be able
to effectively design spacecraft. This has led to the new discipline of
spacecraft–environment interactions. The emphasis in this book is on the
fundamental physics of the interactions.

  Spacecraft–Environment Interactions is a valuable introduction to the
subject for all students and researchers interested in the application of
fluid, gas, plasma, and particle dynamics to spacecraft and for spacecraft
system engineers.

SPACECRAFT—ENVIRONMENT INTERACTIONS

Cambridge atmospheric and space science series

Editors

Alexander J. Dessler
John T. Houghton
Michael J. Rycroft

Titles in print in this series

SPACECRAFT–ENVIRONMENT INTERACTIONS

DANIEL HASTINGS
Massachusetts Institute of Technology

HENRY GARRETT
The California Institute of Technology

CAMBRIDGE
UNIVERSITY PRESS

PUBLISHED BY THE PRESS SYNDICATE OF THE UNIVERSITY OF CAMBRIDGE
The Pitt Building, Trumpington Street, Cambridge, United Kingdom

CAMBRIDGE UNIVERSITY PRESS
The Edinburgh Building, Cambridge CB2 2RU, UK
40 West 20th Street, New York NY 10011–4211, USA
477 Williamstown Road, Port Melbourne, VIC 3207, Australia
Ruiz de Alarcón 13, 28014 Madrid, Spain
Dock House, The Waterfront, Cape Town 8001, South Africa

http://www.cambridge.org

First published 1996
First paperback edition 2004

A catalogue record for this book is available from the British Library

Library of Congress Cataloguing-in-Publication Data

Hastings, Daniel.
Spacecraft-environment interactions/Daniel Hastings, Henry
Garrett.
p.   cm. – (Cambridge atmospheric and space science series)
Includes bibliographical references and index.
ISBN 0 521 47128 1 (hardback)
1. Space environment. 2. Space vehicles – Design and
construction.
I. Garrett, Henry B. II. Title. III. Series.   95-47376
TL1489. H37   1996   CIP
629.4′16 – dc20

ISBN 0 521 47128 1 Hardback
ISBN 0 521 60756 6 Paperback

Contents

Illustrations

Tables

Preface

At the very beginning of the space age, spacecraft designers learned that the effects of the space environment on a spacecraft's systems would be vital factors in spacecraft design and operation. Since those early years, the topic of spacecraft–environment interactions has developed into a multidisciplinary field involving engineers and scientists from all over the world. Traditionally, engineers have been interested in spacecraft design and operational issues, and scientists have concentrated on the fundamental physics and chemistry associated with the interactions. These diverse interests have led to numerous books and conferences. The field has grown substantially in the past decade with the advent of the Shuttle and the ability to perform repeatable, in-situ experiments. The authors therefore concluded that, with the growth of the field and the expanding interest in it, it was timely to prepare a comprehensive book summarizing the many recent discoveries. In particular, since the field has evolved in a way that has been driven by mission and spacecraft requirements rather than as a specific discipline, a book would be a valuable step in integrating the field intellectually. Such a book would also serve as an introduction to the discipline for graduate students and professionals. For specific applications, these individuals could then turn to one of the handbooks or collections of conference papers referenced throughout the book.

This book is the direct outgrowth of courses that the authors have taught. One of us (DEH) has for several years taught a class at the Massachusetts Institute of Technology (MIT) on the subject of spacecraft–environment interactions. This course, aimed at MIT seniors or first-year graduate students, is intended to serve as an entry point for students wishing to go on to research in the field of spacecraft–environment interactions or to learn about the discipline so that they can then go on to spacecraft system engineering. It is also intended as the final course for undergraduate or Master's engineering students interested in the applications of fluid, gas, and plasma dynamics to spacecraft. The other author (HBG), an internationally recognized consultant in the field, has developed a short course based on his actual experiences in spacecraft interactions which has been taught throughout the U.S. Air Force, industry, and internationally. He has participated in almost all of the recent interplanetary missions and many of the Earth-orbiting spacecraft (*SCATHA*, TDRSS, INTELSAT, Galileo, Cassini, HST, *Space Station*, and *Clementine*). These missions have provided numerous practical examples for the book and illustrate many of the issues that arise in designing spacecraft or diagnosing failures.

The central idea developed in this book is that every spacecraft has a set of characteristic interactions (to be defined) with the ambient environment and its self-induced environment. These interactions affect the basic operation of the spacecraft as well as any scientific experiments undertaken from the spacecraft platform. The interactions in turn shed light on the ambient environment. The emphasis in this book is on the physics of the interactions in order to introduce, the basic concepts. However, at the end of most of the sections of the book, the reader will be able to make a simple estimate of the effect of the interaction on the spacecraft. For specific applications, the reader is referred to other books in the field aimed at professional engineers and scientists (also see the extensive References provided):

(1) Al'pert, Ya. L., Gurevich, A. V., and Pitaevskii, L. P., *Space Physics with Artificial Satellites*, Consultants Bureau, New York, 1965

(2) Garrett, H. B., and Pike, C., *Space Systems and Their Interactions with Earth's Space Environment*, Progress in Aeronautics and Astronautics, Vol. 71, AIAA, Washington, DC, 1980

More recent works are:

(3) DeWitt, R. N., Duston, D. P., and Hyder, A. K., *The Behavior of Systems in the Space Environment*, Kluwer Academic Publishers, 1994

(4) Tribble, A., *The Space Environment: Implications for Spacecraft Design*, Princeton University Press, 1995

A book that contains aspects of the interaction issues, but from a systems engineering viewpoint, is:

(5) Wertz, J. R., and Larsen, W. J., *Space Mission Analysis and Design*, Kluwer Academic Publishers, 1991

The chapters in the book are roughly divided between those describing the environment and those describing the interactions. The chapters on the space environment treat both the ambient and the induced environments. The chapters describing the interactions introduce the interaction, describe the relevant physics and chemistry, and then describe qualitatively the practical issues for a spacecraft. Each chapter has extensive references and a bibliography. However, because this book is aimed at individuals entering the field, we have generally included only references that can be found easily in a research library. Occasionally, it is necessary to refer to conference proceedings when no other source of information is available. However, given the difficulties in obtaining copies of the older proceedings, this is kept to a minimum.

The book is arranged as follows.

In Chapter 1, the environments that a spacecraft experiences are described qualitatively. The practice of describing the environment in terms of common orbits is introduced. Next, the notion that the spacecraft is an active player in its environment is discussed. Finally, the history of spacecraft–environment interactions that have been identified and their impact on spacecraft operations are briefly reviewed.

In Chapter 2, the concepts of the characteristic lengths, time scales, and critical velocities for plasma physics, gas dynamics, and radiation physics are introduced. The relevant equations that describe the interactions are introduced along with the terminology and approximations necessary to reduce the equations that describe the physics down to manageable solutions. For example, in the area of plasma physics, the concept of a Debye length and the comparison of this characteristic length to the spacecraft body scale are introduced. The physical approximations that are allowed in the Poisson equation when the Debye length is large or small compared to the body scale will then be presented. In the area of gas dynamics, the Knudsen number is introduced and the physical approximations that result when the Knudsen number is small or large are addressed. For radiation interactions with matter, the idea of linear energy transfer is introduced, which permits straightforward estimates of single-event effects on microelectronic devices.

In Chapter 3, the ambient space environment is described. The ambient environment is analyzed in terms of the regions encountered along common spacecraft orbits or of the type of environment (plasma, radiation, neutral particle, or macroscopic particulates). The intent is to allow the reader to estimate the ambient environment around a spacecraft along a typical orbit.

In Chapter 4, the physics of the interactions of a neutral gas with a spacecraft are addressed. Interactions such as drag, sputtering, glow, and contamination are introduced. The modification of the environment by spacecraft operations is discussed. The physics of each type of neutral interaction are explored using the concepts developed in Chapter 2.

In Chapter 5, the interactions of the space plasma with a spacecraft are addressed. Interactions associated with charging, arcing, electromagnetic interference, and power generation are introduced. The physics of each type of plasma interaction are developed. The operational spacecraft database built upon these interactions is explored along with the implications of each interaction for the design and operation of a typical spacecraft.

In Chapter 6, the interactions of the space radiation environment with spacecraft materials are addressed. Of the interactions, physical damage due to radiation and single-event effects are the most important. Recent information from the *CRRES* satellite are incorporated.

In Chapter 7, the interactions of macroscopic particulates with a spacecraft are discussed. Impact of a spacecraft with orbital debris and scattering of electromagnetic radiation from particulates are introduced. Tactics for designing effective debris and meteoroid shields are outlined and compared with case studies.

In the concluding chapter, Chapter 8, the state of the field as it existed in 1995 is discussed. In addition, future trends in environmental interactions are addressed. A critical issue that emerges in this chapter is the inherent cost of carrying out detailed analyses of the interactions – the means for assessing what interactions are important for a specific mission is an important consideration for the future.

Finally, a word about units. As far as possible we have used MKS units with the choice that energies are frequently in electron volts (eV). However, in the field of environmental interactions, there is still widespread use of CGS units and their use is unavoidable, especially when using figures from older works. The convention that we adopt is to give all formulae in MKS units unless we explicitly state that they are in other units.

Daniel Hastings
Henry Garrett

Acknowledgment

This book is the product of years of research and support from countless individuals. We would like to acknowledge MIT for allowing one of us (DEH) to take time off to write this book. At this time the Air Force Phillips Lab at Hanscom AFB must be acknowledged for their hospitality and for providing an intellectual environment where high-quality work is possible. The support and encouragement of Dr. David Hardy was critical. The stellar work of Robie SamantaRoy, Gabriel Font-Rodriguez, Jim Soldi, David Oh, Carmen Perez de la Cruz, Graeme Shaw, and Derek Plansky in providing pictures as well as insightful comments is also greatly appreciated. Henry Garrett would also like to acknowledge his two mentors in this field: A. J. Dessler (space environments) and N. J. Stevens (space interactions). Many colleagues also provided insightful comments and read the draft. Without them, this work could not have been completed. Particular credit goes to Carolyn Purvis of NASA Lewis and Albert Whittlesey of JPL who read the entire manuscript and have always been great supporters of the field. Finally, we would like to acknowledge our wives, Donna Hastings and Katherine Garrett, and our families whose encouragement and support made this work possible.

1

Introduction

1.1 Introduction

Before the space age began, it was realized that space was not empty. Comet tails, meteors, and other extraterrestrial phenomena demonstrated the presence of a "space environment." Much as an aircraft operates in and interacts with the atmosphere (indeed the air is necessary for lift), so a spacecraft operates in and interacts with this space environment. The environment can, however, limit the operation of the spacecraft and in extreme circumstances lead to its loss. Concern over these adverse environmental effects has created a new technical discipline – spacecraft–environment interactions. The purpose of this text is to describe this new field and introduce the reader to its many different aspects.

Historically, the field of spacecraft–environment interactions has developed primarily as a series of specific engineering responses to each interaction as it was identified. Consider the discovery of the radiation belts and their effects on electronics. This led to the development of radiation shielding and microelectronic-hardening technology. Similarly, in the early seventies, the loss of a spacecraft apparently due to spacecraft charging from the magnetospheric plasma led to intense efforts to understand charge accumulation on surfaces in space and to methods for mitigating the effects. Ultimately, these efforts culminated in the 1979 launch of a dedicated spacecraft, *SCATHA* (Spacecraft Charging at High Altitudes), into a near geosynchronous orbit for studying this interaction. Likewise, in the eighties, certain materials were found to erode rapidly in the low-Earth space environment because of chemical interactions with atomic oxygen. This led to the development of complex ground simulation facilities and to the flight of numerous *Shuttle* experiments aimed at characterizing the phenomenology associated with the erosion. Thus, the study of environment interactions can be characterized largely as a response to problems – it has seldom anticipated them.

The next generation of spacecraft likely will be much longer-lived, more sensitive to environmental effects, and more environmentally active. They will be active in the sense that they may emit particulates, gases, plasma, or possibly radiation (electromagnetic and corpuscular) in sufficient quantities to substantially modify the ambient environment in their vicinity. These spacecraft will possess increasingly more complex, sensitive, and, by inference, expensive instruments. A good example is the *Shuttle*, around which the neutral pressure has often been measured to be over an order of magnitude or larger than the ambient environment. The enhancement is due to outgassing, water releases, and thruster firings. It and similar self-generated environments may significantly alter the radiation and plasma components. These in turn could pose serious problems for the operation of sensitive optical and electrical/electromagnetic sensors or threaten the long-term integrity of the spacecraft structures and electronic systems. The greatly increased cost of future systems such as the *space station* will require a long operational lifetime to amortize the costs. This may mean that even seemingly innocuous interactions could, through cumulative effects, reduce the lifetime of the system rendering it uneconomical or unfeasible. No longer will the space engineer have the luxury of fixing problems after the fact; they must be anticipated in the original design.

In addition, for the new class of vehicles, many of the interactions may be synergistic, greatly enhancing their impact. That is, relatively weak individual interactions could couple in such a way as to have a nonlinear effect on the spacecraft, becoming strong enough to be design limiting. For example, the choice of a negative ground for the high-voltage power system on the *space station* may increase the probability of arcing on the structure. This arcing may erode thermal control coatings and increase the contamination in and near the station. For this reason, a plasma contactor has been incorporated into the station design to eliminate such arcing. To understand and control such environmental interactions, it has become critically important to develop a unified description of the spacecraft, the environment, and the interactions. That description forms the basis of a new engineering and scientific discipline – spacecraft–environment interactions.

Since interactions and their effects can depend on the environment, the spacecraft, and the spacecraft subsystems, it is important to properly define these variables. Based on these definitions, it is possible to systematically organize and describe the basic interactions. This procedure is followed in the next section.

1.2 Classification of Spacecraft Environments

The environment to which a spacecraft is subject consists of the combination of the ambient (typically a function of the orbit) and that generated by the spacecraft itself. The combination of these environments may not be their simple sum but a more complex environment brought about by a synergistic, nonlinear interaction. In fact, the self-generated environment of a spacecraft may substantially differ from the ambient, suggesting that the orbit may not always be a primary consideration in characterizing the in-situ spacecraft environment. In any event, in this book, the term "spacecraft environment" always means the combination of the ambient and the induced environments.

It is useful to characterize the environment in terms of four physical components: the neutral environment, the plasma environment, the radiation environment, and the particulate environment. The neutral environment includes the ambient gas and that released by the spacecraft surface materials through outgassing or decomposition, deliberately vented from the spacecraft, or emitted during thruster firings. The plasma environment includes the ambient plasma; that released from plasma thrusters; that created by ionization of or charge exchange with, the neutral gas; that generated by arc discharges; or that created by hypervelocity impacts with the spacecraft surfaces. The radiation environment has two components: electromagnetic and corpuscular. The electromagnetic radiation environment includes the ambient solar photon flux, that reflected (and emitted) from the Earth, and the electromagnetic interference (EMI) generated by the operation of spacecraft systems or arcing. It also includes electromagnetic waves generated by the plasma environment and photons emitted from spacecraft nuclear sources. The corpuscular radiation environment consists of the ambient flux of particles (electrons, protons, heavy ions, and neutrons) and any high-energy particles emitted by nuclear sources or reactors. The particulate environment consists of ambient meteoroids, orbital debris, and particulates released by the spacecraft. These are from a number of sources ranging from dust on the surfaces to material decomposition under thermal cycling and exposure to ultraviolet radiation.

1.3 Spacecraft Orbits and the Ambient Space Environment

Spacecraft orbits typically fall into specific families based on the intended use of the spacecraft. In addition to defining the interactions in terms of specific environmental conditions, it is therefore useful to consider the cumulative effects along these common orbital paths. There are five families of orbits that are of particular relevance for spacecraft interactions near the Earth. Other planets have the same components to the environment but different characteristics. These are: low Earth orbit (LEO), medium Earth orbit (MEO), polar orbit (PEO), geosynchronous orbit (GEO), and interplanetary orbit. Although a given spacecraft mission may have a more complex trajectory than represented by these orbits, it is still common to refer to the interactions that the spacecraft will see in terms of the five families. The characteristics of the five orbits are listed in Table 1.1.

Table 1.1. *Classification of orbits*

Name	Altitude (km)	Inclination to equator (deg)
Low Earth orbit	100–1,000	<65
Medium Earth orbit	1000–36,000	<65
Polar orbit	>100	>65
Geostationary orbit	~36,000	0
Interplanetary orbits	Outside magnetosphere	N/A

Table 1.2. *Description of orbits*

Name	Description
Low Earth orbit	Cold, dense, ionospheric plasma; dense, supersonic neutral atmosphere; solar ultraviolet (uv); orbital debris; South Atlantic anomaly (SAA)
Medium Earth orbit	Solar uv; trapped radiation belts; plasmasphere
Polar orbit	Solar uv; cold, dense ionosphere; supersonic neutral atmosphere; orbital debris; auroral particles; solar flares; cosmic rays; SAA; horns of radiation belt
Geosynchronous orbit	High-energy plasmasheet; substorm plasma; uv radiation; outer radiation belts; solar flares; cosmic rays
Interplanetary orbits	Solar-wind plasma; solar flares; cosmic rays

The primary physical components of the environment associated with each of the orbit families are described qualitatively in Table 1.2 (quantitative values are given in later sections). As an example of this classification scheme, consider a nominal *space station (SS)* orbit of 28.5° and *Earth Observing System (EOS)* satellite orbit. Their orbits are described in Table 1.3 and can be classified as being affected by the LEO or LEO/PEO orbital environments, respectively. Of course, there are highly elliptical orbits that span all five orbital environments. In such cases, a designer has to consider the characteristic interactions for each

Table 1.3. *Assumed* Space Station *(SS) and* Earth
Observing System *(EOS) orbits*

	Spacecraft	
Orbit	*SS*	*EOS*
Inclination (deg)	28.5	98.25
Altitude (km)		
Minimum	463	400
Nominal	500	705
Maximum	555	900
Orbit type	LEO	LEO/PEO

orbital segment as the vehicle passes through the different orbital regions along its trajectory.

1.4 Spacecraft Systems

Spacecraft require many different types of systems for their successful operation. Each system may affect or be affected by the environment. The systems also may add to the induced environment around the spacecraft. Typical systems are: power, propulsion, attitude control, structure, thermal control, avionics, communications, and the payload. A brief description of each of these components follows (for a complete description of spacecraft systems, see Agrawal (1986) and similar references).

Power System: The power system provides the electrical power for the spacecraft and its payload. For spacecraft with orbits inside the asteroid belts, the power source is usually solar arrays, although it can be fuel cells as on the *Shuttle*. For missions to the outer planets, nuclear sources such as radioisotope thermal generators (RTGs) are required. The power system also includes the power processing units and the power distribution subsystem (i.e., the cables, relays, and electronics necessary to get the power to where it is to be used) and the power storage system (usually batteries).

Propulsion System: The propulsion system is responsible for providing the velocity increments (or Δv) needed to maneuver and boost or reboost the spacecraft. The propulsion system is generally the chemical or plasma thrusters along with the associated tanks, propellant, and plumbing.

Attitude Control System: The attitude control system senses the spacecraft orientation relative to some reference system (e.g., the Earth, fixed stars, or the Sun) and maintains a desired attitude. It is composed of sensors such as star trackers or horizon sensors and inertial measurement units (IMUs) and actuators such as control moment gyroscopes, control thrusters, magnetic torquers, and flywheels.

Structure: The spacecraft structure physically houses all the systems of the spacecraft and includes the internal structure (e.g., plates, decks), external appendages, and the surface materials that make up the spacecraft skin. For understanding interactions with the environment, this book is concerned mainly with the external structure and the spacecraft skin. The structure or some part of it is usually taken as the electrical reference (i.e., the spacecraft ground).

Thermal Control System: The thermal control system is responsible for maintaining the temperature of the spacecraft within acceptable limits. It can be active, passive, or some combination of the two. A typical system is composed of heaters, coolers, radiating surfaces, and means for conducting heat around the spacecraft. Examples of the latter are heat sinks or heat pipes. Surface materials are often selected for their thermal properties, and thermal blankets or coatings frequently dominate the spacecraft exterior surfaces.

Avionics System: The avionics system has the task of controlling the functions of all the other systems and operating the spacecraft. It is composed of the electronics as well as the software necessary to run the spacecraft.

Communications System: The communications system provides the two-way command and data relay link with the ground station. It is composed of the transmitters, receivers, spacecraft antennas, and actuators necessary to orient them.

Payload: The spacecraft payload typically has many functions. For the purpose of studying interactions, however, the major payload components considered will be limited to different types of sensors and communications devices.

1.5 Interactions between the Environment and a Spacecraft

In this section, as an overview, the effects of the environmental components are summarized. Each of the four environmental components can affect the design and operation of a space vehicle or its systems. The effects may not be constant over time and will often change as the vehicle ages. Even on very short time scales (a fraction of the orbital period), the effects of an environmental interaction can vary substantially. In addition, although each environmental component has a unique effect on the spacecraft, it will be instructive to group their effects in terms of the five orbit families.

Consider first the neutral gas environment. This component has a number of potentially adverse effects on spacecraft. The ambient neutral environment in LEO below $\sim$800 km is dominated by the Earth's residual atmosphere, which is primarily monatomic oxygen over most of the altitude range (see Section 3.2). The atmosphere exerts an aerodynamic drag force on spacecraft. This drag force arises from the impact of the atmospheric particles on the spacecraft surfaces. Although the drag force is typically antiparallel to the spacecraft velocity vector, for large asymmetric spacecraft, aerodynamic torques become an issue. They must be taken into account by the attitude control system and can cause long-term problems for a large vehicle such as the *space station*. For LEO, the aerodynamic drag will eventually deorbit the spacecraft if it is not countered by periodic reboosting of the spacecraft. For example, the *space station* will require one logistics *Shuttle* flight a year to replace the propulsion modules and keep the station in orbit.

The impact of the atmospheric molecules on the spacecraft in LEO can initiate physical and chemical changes to the materials making up the structure of the space-craft. The mean kinetic impact energy of the dominant atomic oxygen impinging on frontal or ram surfaces is 5 eV. Although generally this is not energetic enough to physically remove material from the surface, it is energetic enough to initiate chem-ical reactions on certain materials at the spacecraft surface that can lead to material loss. The flux of atomic oxygen to spacecraft surfaces at low-Earth orbital condi-tions (with a speed relative to the ambient atmosphere in the 7- to 8-km/s range) is approximately a monolayer per s. This flux has been shown to lead to surface erosion of materials such as Kapton or silver. For example, unprotected Kapton, a material often used as an external thermal control surface, was completely eroded from exposed surfaces on the Long Duration Exposure Facility (*LDEF*) spacecraft. The *LDEF* was placed in LEO orbit by the *Shuttle* and orbited for six years before being retrieved. Even on the short *Shuttle* missions, exposed Kapton samples have been found to erode measurably in a few days. In another example, one of the early designs of the *space station* was to have used a carbon–carbon composite for the truss as a mass-saving material relative to aluminum. It was determined that such a composite would erode significantly after only five years in space. Even if the flux of atomic oxygen does not erode the surface, oxidation of the surface may change the thermal properties of the surface layer. This must be considered in the design of the spacecraft thermal control system.

The LEO ambient neutral component is also a direct contributor to the diffuse UV–visible–IR glows that have been observed to occur above surfaces oriented to-ward the spacecraft ram direction. These complex glow phenomena, which include surface-catalyzed, excited recombination, appear to be functions of the spacecraft altitude, attitude, materials, surface temperature, time in orbit, nature of the orbit (including sunlight conditions), and vehicle size.

The induced neutral environment around a spacecraft arises from the release of neutral gas from sources on the spacecraft. Many materials are known to release absorbed gas on exposure to the space environment because the ambient neutral gas pressure in space is so low relative to that of the Earth. Additionally, materials may release gas through decomposition or sublimation. Neutral gas is generated through backflow from thruster firings, incomplete ionization of ion thruster gases, and effluent dumps. Over time, these gaseous products can coat and contaminate sensitive sensors and surfaces, seriously degrading their performance or rendering them useless. Optical sensors may be affected on the payload as well as thermal control surfaces and coverslides on solar arrays. One example of how interactions with supposedly neutral gases can drive the spacecraft design is given by the *Hubble Space Telescope* (*HST*). For the *HST*, the desire to protect the mirror from contamination led to the decision not to place attitude control thrusters on the spacecraft. Instead, the attitude is controlled by momentum wheels and magnetic torquers.

The plasma component of the environment represents a current flow to the spacecraft skin and the exposed parts of the power subsystem. Intrinsic imbalances in this current flow result in the buildup of charge on all surfaces exposed to the plasma. Charging also can be caused by the photoelectric effect, which causes surfaces to emit low-energy electrons when they are illuminated by the Sun. For large spacecraft in LEO, currents may be induced by the motion of the spacecraft across the geomagnetic field. The current flow to the spacecraft also may be significantly modified by the electric fields generated by a high-voltage power system exposed to the space environment.

The effects of current flow to the spacecraft can be profound because it can cause differential charge accumulation on the spacecraft surfaces. This charge, in turn, produces potential gradients between electrically isolated surfaces of the spacecraft and relative to the spacecraft ground and space plasma. At a minimum, any shift in potential relative to the spacecraft ground or to the space plasma can affect the operation of instruments designed to collect or emit charged particles. Beyond that, the buildup of differential potentials on the surface of the spacecraft or on the power system can give rise to destructive arc discharges or microarcs that generate electromagnetic noise and erode surfaces. This surface erosion contributes to the gas and dust environments near the spacecraft. For highly biased solar arrays (generating greater than 1,000 volts), it has been found that the arcing induced by the LEO plasma for conventionally designed solar cells is so severe that it destroys the array. Even for much lower voltages, the desire to avoid microarcs and the associated electromagnetic interference has been a design-limiting factor for solar arrays. Indeed, the *space station* solar arrays were chosen to operate at 160 volts to stay comfortably below an empirically determined arcing threshold of 200 volts. This lower voltage increased the weight of the power distribution compared to that for the higher operating voltages originally envisioned (higher voltages equate to lower line losses for the same thickness of wire).

For spacecraft in GEO, the charging environment can be much more severe than in LEO because the plasma, though much more tenuous, is very energetic.

This plasma can sustain surface potential differences of several thousands of volts between the spacecraft structure, its surfaces, and the space plasma. The arcing associated with the appearance of such large potential differences is believed to have been directly responsible for the failure of at least one and perhaps several GEO satellites as well as anomalous behavior on many others. To mitigate the effects of the charging, detailed design guidelines and computer codes have been developed by NASA to determine the type and placement of materials on the spacecraft surface, grounding schemes, and circuit filters. Although necessary, the synergistic relationship between these charge control design considerations, the thermal control design, and, in some cases, the meteoroid protection system can greatly complicate the design of a spacecraft.

One active way to mitigate the effects of surface charging is to emit a dense, cold plasma from a source on the spacecraft. The dense plasma supplies the charge required to neutralize the differential charge buildup on the surface and to balance the currents due to the ambient plasma while maintaining a desired frame potential. This technique was successfully demonstrated on the Advanced Technology Satellite, *ATS-6*, and will be used on the *space station* to suppress arcing on the habitation module. Besides arcing, the charge buildup on the spacecraft can attract charged contaminants to sensitive surfaces. This contamination, in turn, can alter the properties of the surface (e.g., making a conducting surface less conducting) and change the charging characteristics. This is known to occur on satellites at GEO and is one example where the synergism of the plasma and the neutral environment can produce an effect that enhances both interactions. Another example is that the flux of these neutral species to a surface can enhance the possibility of arcing associated with exposed parts of the power system by providing a source of electrons through ionization (i.e., Paschen breakdown or multipacting).

The corpuscular (particle) radiation component of the environment can affect the vehicle systems by direct radiation damage as well as by deep dielectric charging. The latter process is the result of high-energy electrons that can penetrate into the interior of a vehicle, deposit charge, and, ultimately, induce arcs inside the vehicle on electrically isolated components. The direct radiation damage can be either temporary or permanent. Temporary damage occurs when the state of an electronic component is momentarily modified by the passage of a high-energy particle through the component. This is known as a single-event effect (SEEs) and can reset a spacecraft clock, change the state of a random access memory, increase the noise levels in charge-coupled devices, and induce other false signals. In particularly severe cases, the SEE can cause a 'latchup,' where permanent damage can result from burnout of the integrated circuit. More common interactions are associated with the long-term buildup of the total ionizing dose (TID). The slow accumulation of charge or physical damage to the material because of the passage of high-energy particles leads to power loss in solar cells, degradation and failure of microelectronics, and darkening of optical components. Indeed, radiation damage to solar cells is one of the most important life-limiting factors in power system design for spacecraft. The design solution of choice is to oversize the solar array so that it will still be producing the desired power levels at the end of the life of the spacecraft. Clearly, such a solution is wasteful of weight and hence costly for the spacecraft program. Proper design to protect the spacecraft and its systems from the effects of corpuscular radiation can be extremely expensive, particularly for the new, more susceptible microelectronic components coming on the market. It is a major driver in the study of spacecraft interactions.

In addition to the effects of radiation due to particles, photon radiation effects also can adversely affect spacecraft systems. At the lowest frequencies, radio frequency interference affects the electronic systems while infrared from the Earth or other celestial body can alter the thermal balance of an orbiting spacecraft. Visible light glinting off surfaces or dust in the vicinity of sensors can create false images. At the other end of the frequency band, the ultraviolet radiation environment in space can directly degrade the properties of many of the materials used on spacecraft surfaces. As mentioned before, it can modify the charging of a spacecraft through photoemission or by photochemically bonding contaminants to sensitive surfaces. X rays and gamma rays, primarily from man-made sources, can penetrate surfaces and generate charged particles inside the spacecraft shielding, greatly enhancing their effect on sensitive systems.

Finally, impacts by the meteoroid or space-debris particulate environments can damage or totally destroy a spacecraft. The kinetic energy of even small particles moving at low-Earth-orbital velocities is so large that severe damage can result (the impact of an object the size of a pea moving at low-Earth-orbital velocities of 7 km/s is comparable to a bowling ball moving at 60 mph). Micrometeoroid impact velocities are typically 15 to 20 km/s and can be as large as 70 km/s. For example, a paint fleck struck a *Shuttle* window with sufficient energy to cause a large enough pit to require window replacement. The issue of damage from orbital debris has become one of the driving issues for the design of the *space station*. In addition, impacts can induce arcing on structures and surface materials under voltage stresses of less than 100 volts.

Besides the obvious potential for damage from particles moving at a large relative velocity with respect to the spacecraft, near-field particulate contamination can seriously degrade the performance of spaceborne optical systems. Small particulates trapped near the vehicle radiate or scatter enough energy to overload sensitive sensor systems. A particle in the near field that radiates may exceed the signatures of targets that are in the far field. Consequently, the particles will appear as clutter in the field of view of the sensor system. Dust or particulates around a spacecraft

Table 1.4. *Impact of spacecraft environments on spacecraft systems*

Spacecraft system	Spacecraft environments			
	Neutrals	Plasma	Radiation	Particulates
Power	Change in coverglass transmittance	Shift ground, attract contaminants, arc damage	Degradation of solar cell output, arc damage	Destruction of solar cells
Propulsion	Contaminant source, drag	Contaminant source		Particulate source
Attitude control	Torques, sensor degradation	Torques	Sensor degradation	
Structure	Erosion	Arc damage	Arc damage	Penetration
Thermal control	Change in surface properties	Change in surface properties	Change in surface properties	
Avionics		EMI	Degradation	
Communication systems		EMI		
Payload	Sensor interference (e.g., glow)	Sensor interference	Avionics damage	Penetration

can be created by flaking of the skin of the spacecraft, firing of rocket motors (especially solid rocket motors), and dumps of effluents which form crystals in the space environment. The control of these particulates, which become space debris on departure of the source vehicle, is proving very difficult. Their near-exponential growth at LEO is beginning to threaten the long-term viability of this regime for space operations, particularly manned operations!

All of these interactions are summarized in Table 1.4, which lists their impacts by spacecraft subsystem (after Tribble, 1993).

1.6 Historical Review of Spacecraft–Environment Interactions

1.6.1 The Dawn of the Space Age to 1970

The history of spacecraft–environment interactions closely follows the history of the space age itself. The modern space age is usually assumed to have started with the launch of *Sputnik* in 1957. If so, then the discovery in 1962 by Van Allen and his collaborators on *Explorer 1* of the existence of toroidal belts of energetic protons and electrons encircling the Earth can be said to be the beginning of the study of spacecraft interactions. This discovery led to many missions to study these radiation belts, with the result that, by the early seventies, several models of the radiation belts and their effects had been successfully developed. There was also a growing appreciation of the subtle interactions between a spacecraft, the neutral atmosphere (principally through drag effects), and the ionospheric plasma. This appreciation was driven both by scientific interest and by the need to have the spacecraft and its payloads perform optimally in this environment. Indeed, much of the fundamental physics of spacecraft interactions (as then known) was defined and understood by the beginning of the seventies. This period, in particular, produced some excellent books on the aerodynamics of flight through the ionosphere (Al'pert, Gurevich, and Pitaevskii, 1965; Singer, 1965; Kasha, 1969).

In addition to these direct interactions, more indirect interactions due primarily to contamination also were recognized. Effects on many spacecraft from outgassing and particulate contamination were noted. For example, on a *Mariner* mission, there was loss of the star tracker lock that was traced to particulate contaminants in the spacecraft near-field. On the *Gemini* missions, the windows became contaminated because of the use of silicones in the window gaskets. In the *Apollo* missions, high background signals were observed by mass spectrometers. This enhanced background was traced to outgassing from the spacecraft. Thus, by the beginning of the seventies, radiation effects, micrometeoroids, plasma interactions (primarily at low altitudes), and contamination were recognized as legitimate design concerns – models and mitigation techniques were available. This set the stage for the discovery and investigation of an entirely new family of interactions in the seventies.

1.6.2 The Decade of the Seventies

The seventies saw attention shift to spacecraft charging and its effects as observed on GEO spacecraft (GEO was becoming a region of increasing commercial importance). Anomalies were observed on several GEO spacecraft (Rosen, 1976) that were believed to correlate with surface charging. A vigorous research program was launched in the United States as a joint effort between the Air Force and NASA. It consisted of laboratory work on material properties and charging processes, initiation of the development of a large-scale simulation code (the NASA Charging and Analysis Program, or NASCAP), and the launch of a dedicated spacecraft in 1979, called the *SCATHA* satellite, to study the phenomena. In the USSR, there was a similar effort that resulted in the development of two simulation codes akin to NASCAP. The principal book that summed up the state of the art in spacecraft charging and spacecraft–environment interactions in general up to 1980 was the volume by Garrett and Pike (1980). This book introduced the whole range of environmental effects and, for the first time, addressed the concept of mankind's permanent impact on the space environment. It also included a reprint of one of the first papers on space debris (Kessler and Cour-Palais, 1980), an issue that would come to dominate future interaction studies.

1.6.3 The Decade of the Eighties

The eighties saw the culmination of the GEO charging program with the release of NASCAP, publication of the *SCATHA* results (Koons, 1983), and establishment of a GEO spacecraft anomalies database. The codes and associated design procedures to alleviate spacecraft charging were incorporated by NASA into a comprehensive design guidelines document that is still widely used throughout the community for spacecraft design (Purvis et al., 1984).

With the dawn of the *Space Shuttle* era, attention shifted from GEO to effects important to systems in LEO. The plasma-effects community focused attention on high voltages and active systems (e.g., electrodynamic tethers and the *space station*) interacting with the ionosphere. In 1983, following initial indications from other LEO spacecraft, scientists searched for light emissions in the vicinity of the *Shuttle* during the flight of STS-3 (Banks et al., 1983). The *Shuttle* surface was observed to glow brightly and select materials were found to erode rapidly in the "benign" LEO environment. These observations led to a substantial ground- and space-based effort to characterize these phenomena. Many ground facilities were developed to simulate the LEO neutral environment. This proved difficult, however, given the nature of this environment – primarily a high flux of ground-state atomic oxygen moving at 7 to 8 km/s. On the other hand, the *LDEF* provided a wealth of in-situ data on material effects during its six years of long-term exposure to the LEO space environment. Taken together, these studies led to major strides in the understanding of hypervelocity interactions between the upper atmosphere and spacecraft.

Orbital debris was recognized early in the eighties as a significant hazard to LEO vehicles. Efforts to assess, control, and mitigate this new environment were initiated (Johnson and McKnight, 1991). Institution of the Strategic Defense Initiative Organization program in the middle of the decade focused interest on very-high-power system effects and radiation damage to microelectronics (required for the "brilliant" weapons that were planned). Indeed, by the end of the eighties, attention had shifted firmly to large, high-powered vehicles such as the *space station*. It was realized that vehicles such as the *space station* or the Star Wars weapons introduced new classes of interactions both because of the increased synergy of the interactions and because they operate in environmental-effects regimes that have not yet been explored. This increasing interest in the effects associated with large, high-voltage systems indeed will likely be a major theme for the next decade. Given the exceptionally high cost of these large systems and their required lifetimes (10 to 30 years), even low-level interactions could be critical.

1.6.4 The Nineties

What is the future for spacecraft–environment interactions? We believe that the nineties offer new vistas in the field of spacecraft interactions. In addition to the *space station*, three other new classes of spacecraft are likely to become important. Each of these will offer interesting environmental interaction challenges. The first of these are spacecraft propelled by plasma thrusters. Plasma thrusters substantially alter the plasma environment in their vicinity. Since they will be operating for a long time, a spacecraft will develop its own, unique steady-state plasma environment. The second class is that of spacecraft designed to operate continuously in MEO as opposed to just passing through the radiation belts periodically. These will have to be designed to be unusually radiation hardened and fault tolerant. The third class will be small spacecraft designed for low-cost launch and operation rather than ultrareliable performance. This change in the design driver may have profound effects on spacecraft design as engineers attempt to keep down costs and to minimize the effects of interactions – potentially contradictory goals. Thus the nineties will offer many new challenges!

1.7 Purpose of the Book

This chapter presents a qualitative description of the interactions of a spacecraft with its environment. These interactions may be design limiting, as in the case of arcing on high-voltage solar arrays, or limit operational life, as in the case of radiation damage to solar cells. Most types of interactions are merely annoyances, however. An example in this latter class is the temporary loss of star tracker lock because of near-field particulate contamination. Although this can be dealt with easily by instructing the star tracker to ignore some number of uncorrelated targets, it has the potential to be extremely annoying, as one of the authors can attest from personal experience. As can be seen from the brief historical review, the field of space environment interactions has been created in large part in response to discoveries of new environments in space or to the identification of new effects on space systems. Although this has led to solutions to the problems raised by the specific interactions, the results have often been quick fixes rather than long-term, well-thought-out system solutions. It is the thesis of this book that a unified and fundamental approach to the discipline of spacecraft–environment interactions is both necessary and timely. Such an approach will enable better engineering solutions to the problems raised by interactions and a better understanding of the physics associated with spacecraft operating in the space environment. In the following chapters, a systematic, unified approach to spacecraft interactions is developed by examining the fundamental physics associated with the four different components of the environment. The objective is to introduce readers to the fundamentals and then to point them to the rich body of literature and data that exist for this interesting and important new discipline.

2

Fundamental Length, Time, and Velocity Scales

In Chapter 1, the four basic environmental interactions were introduced. To help in understanding the physics of these interactions with a spacecraft, the nondimensional physical parameters that determine the interactions are described in Chapter 2. An understanding of the magnitudes of these nondimensional parameters simplifies the complex physics describing the interactions. This is analogous to the idea that, for a fluid, the definition of a Reynolds number allows the physics of the fluid behavior to be divided into two regimes. When the Reynolds number is small compared to unity, the physics of the fluid is dominated by viscous effects. When the Reynolds number is large compared to unity, the flow is inviscid. Likewise, for a compressible gas, the Mach number allows the physics of the flow to be divided into subsonic physics and supersonic physics. This idea of fundamental scales, like the Reynolds and Mach numbers, is exploited in this and subsequent chapters as a means of characterizing the effects and importance of the fundamental interactions under varying environmental constraints.

2.1 The Concept of a Distribution Function

An important concept necessary for establishing scales in the space environment is that the four environmental groups can each be described in terms of a distribution function. The concept of a distribution function, as described by Vincenti and Kruger (1965) and Bittencourt (1986), is that any distribution of particles that is nonuniformly distributed in space can be described by a local number density $n(\vec{x})$ defined at each point in space $\vec{x}$. This function is defined by the following limiting process. If there are ΔN particles contained in the differential element of volume $\Delta V = \Delta x_1 \Delta x_2 \Delta x_3$ where x_i is the ith element of the position vector $\vec{x}$ then

$$n(\vec{x}) = \lim_{\Delta V \to 0} \frac{\Delta N}{\Delta V}. \tag{2.1}$$

16

This limit exists and is meaningful as long as, in the limiting process, ΔV approaches zero on a scale that is large compared with the spacing between particles, such that there is always a large number of particles in the volume. For an ordinary gas, the product $m_p n(\vec{x})$, where m_p is the mass per gas molecule, gives the well-known mass density. This approach shows that $n(\vec{x})$ is the number of particles per unit volume as a function of spatial location. The density $n(\vec{x})$ is thus a measure of how the particles are distributed in real space. Therefore, $n(\vec{x})$ can be called the position distribution function of the particles. The total number of particles in volume V is then given by

$$N = \int_V n(\vec{x})d^3x. \tag{2.2}$$

The integration is a three-dimensional integral over V where the particles can be found and the notation d^3x denotes the differential volume element $dx_1dx_2dx_3$. The number of particles that can be found in the differential volume element d^3x is $n(\vec{x})d^3x$.

More generally, to completely describe a large set of particles, it is necessary to define the density distribution of the particles in terms of all of the relevant (abstract) spaces associated with the properties of the particles. As a minimum, a particle will have three position coordinates in real space and three vector velocities associated with it. To completely describe the particle, therefore, the abstract velocity space is introduced. In this space, the velocity vector, $\vec{v}$, is a vector from the origin to the point in the space described by the coordinates $(v_{x_1}, v_{x_2}, v_{x_3})$ where v_{x_i} is the velocity in the direction x_i. The instantaneous state of any particle is defined by the independent coordinates $x_1, x_2, x_3, v_{x_1}, v_{x_2}, v_{x_3}$. This fact suggests the idea of a six-dimensional phase space defined by the six coordinates $x_1, x_2, x_3, v_{x_1}, v_{x_2}, v_{x_3}$. Each particle is then represented by a point in this space. In the same manner as discussed above for the mass density, a density of particles can be defined in the six-dimensional phase space. This phase-space density is called the distribution function, f_j, of the particles and is defined as follows. If dN_j is the number of particles of type j that exist in the six-dimensional volume element $d^3x d^3v_x$ at time t, then

$$dN_j = f_j(\vec{x}, \vec{v}_x, t)d^3x d^3v_x. \tag{2.3}$$

It is clear from this definition that f_j is the density of particles in the six-dimensional phase space. It is also clear that since f_j is a density, then it must always be positive and finite. From Eqs. (2.2) and (2.3),

$$N_j = \int_V n_j(\vec{x}, t)d^3x = \int_V \int_{-\infty}^{\infty} f_j(\vec{x}, \vec{v}_x, t)d^3x d^3v_x, \tag{2.4}$$

where the integration is over all of the velocity space. Therefore, it must be the case that

$$n_j(\vec{x}, t) = \int_{-\infty}^{\infty} f_j(\vec{x}, \vec{v}_x, t) d^3 v_x. \tag{2.5}$$

Furthermore, since the density in real space is a well-defined quantity, the function f_j must tend to zero as the velocity becomes very large (in other words, particles with infinite velocities cannot exist).

Equation (2.5) shows that the integral over the velocity space of the distribution function is the macroscopic density. More generally, a moment of $O(K)$ of the distribution function is defined as

$$M_{x_i x_j \dots x_l}^{(K)}(\vec{x}, t) = \int_{-\infty}^{\infty} \underbrace{v_{x_i} v_{x_j} \cdots v_{x_l}}_{K \text{ times}} f(\vec{x}, \vec{v}, t) d^3 v_x. \tag{2.6}$$

The zeroth moment is just the number density as given in Eq. (2.5).

It is also possible to define the average value of any property $\vec{\alpha}(\vec{x}, \vec{v}, t)$ of the particles as follows:

$$\langle \vec{\alpha}(\vec{x}, t) \rangle = \frac{1}{n(\vec{x}, t)} \int_{-\infty}^{\infty} \vec{\alpha}(\vec{x}, \vec{v}, t) f(\vec{x}, \vec{v}, t) d^3 v_x. \tag{2.7}$$

The average macroscopic velocity of the particles is then defined as

$$\vec{u}(\vec{x}, t) = \langle \vec{v} \rangle. \tag{2.8}$$

This definition allows the random velocity of a given particle to be defined as

$$\vec{c} = \vec{v} - \vec{u}. \tag{2.9}$$

Clearly, $\langle \vec{c} \rangle = 0$. This justifies the term *random velocity*.

The first moment of the distribution function is just

$$M_{x_i}^{(1)} = n_j(\vec{x}, t) \langle v_{x_i} \rangle.$$

The second moment is related to the momentum flow tensor, which is defined by

$$P_{ij} = m_p n(\vec{x}, t) \langle v_{x_i} v_{x_j} \rangle = m_p M_{x_i x_j}^{(2)}.$$

The ijth component of the pressure tensor is defined as

$$p_{ij} = m_p n(\vec{x}, t) \langle c_{x_i} c_{x_j} \rangle.$$

Therefore, the momentum flow tensor is related to the pressure tensor by

$$p_{ij} = P_{ij} - m_p n(\vec{x}, t) u_{x_i} u_{x_j}.$$

The scalar pressure is defined by

$$p = \frac{1}{3} \sum_{i=1}^{i=3} p_{ii}$$

and, with the definition $c^2 = c_{x_1}^2 + c_{x_2}^2 + c_{x_3}^2$, is given by

$$p = m_p n(\vec{x}, t) \langle c^2 \rangle / 3. \tag{2.10}$$

The mean kinetic energy per particle is defined as

$$\langle E \rangle = m_p \langle c^2 \rangle / 2. \tag{2.11}$$

When the particles are in or near a situation of thermodynamic equilibrium, a characteristic temperature, T, of the set of particles can be defined. From classic thermodynamics and statistical mechanics, there is an energy $kT/2$ associated with each translational degree of freedom of the particles, so that

$$kT = m_p \langle c_{x_1}^2 \rangle / 2 = m_p \langle c_{x_2}^2 \rangle / 2 = m_p \langle c_{x_3}^2 \rangle / 2, \tag{2.12}$$

where k is the Boltzmann constant. From Eqs. (2.11) and (2.12), it can be seen that $\langle E \rangle = 3kT/2$, and from Eq. (2.10), it can be seen that $p = nkT$. This is the equation of state for an ideal gas.

The concept of a density or distribution function in an abstract phase space is capable of much greater generality than the definition given above. In particular, if the particles of type j have another property or characteristic that may vary from one particle or location to another, then the density can be defined in the space of that property. For example, particles may have different masses or different ionization states. Therefore, an abstract mass space can be defined where each particle will be a point in that space. Similarly, an abstract ionization state space can be defined where each particle is a point in the space. A good example is the distribution of orbital debris and interplanetary meteors. There is a very large range of different particles with different masses that make up orbital debris. They can be described by a distribution function, f_{part}, defined over the velocity and mass space such that the number of particles of orbital debris is

$$N_{\text{part}}(t) = \int_0^\infty dm \int_V d^3x \int_{-\infty}^\infty d^3v_x \, f_{\text{part}}(\vec{x}, \vec{v}, m, t),$$

where the mass integration is over all positive masses.

2.1.1 The Maxwellian Equilibrium Distribution Function

In books on statistical mechanics, gas dynamics, and plasma physics [see Vincenti and Kruger (1965) and Bittencourt (1986)], it is shown that for a system in thermodynamic equilibrium there is a most probable distribution function. This is the well-known Maxwellian equilibrium distribution function and is defined by

$$f_M(c) = n\left(\frac{m_p}{2\pi kT}\right)^{3/2} \exp(-m_p c^2/2kT), \qquad (2.13)$$

where n is the uniform density of the gas and T is the characteristic temperature. Generally, in the ambient space environment and, in particular, for spacecraft–environment interactions (since the spacecraft itself always disturbs the equilibrium), the neutral gas or plasma is not in thermodynamic equilibrium. However, they may be close to equilibrium in the sense that the perturbations to the equilibrium state are small (Nicolis and Prigogine, 1977). In this case, the distribution function of the system can be described by a *local* Maxwellian distribution function,

$$f_M(\vec{x}, \vec{v}, t) = n(\vec{x}, t)\left[\frac{m_p}{2\pi kT(\vec{x}, t)}\right]^{3/2} \exp[-m_p|\vec{v} - \vec{u}(\vec{x}, t)|^2/2kT(\vec{x}, t)], \quad (2.14)$$

where the density n, mean velocity $\vec{u}$, and temperature T are slowly varying functions of space and time.

The Maxwellian distribution function is such a common distribution function in gas and plasma dynamics that enumerating some of its properties is worthwhile. Statistical mechanics, for example, proves that the Maxwellian is the most probable microscopic distribution function satisfying the macroscopic constraints of density conservation, energy conservation, and momentum conservation imposed on a system of particles. It is the distribution function that maximizes the entropy of the set of gas or plasma particles and therefore is the only distribution function that is consistent with the three laws of thermodynamics. It is also the distribution function that any initial distribution will evolve toward as long as there are physical processes present that create randomness in the set of gas or plasma particles. For example, in the ionosphere (see Section 3.3) the number of collisions of particles is sufficiently large that the distribution function would be expected to be close to a Maxwellian. However, in a GEO environment, the collision rate of particles is so low that there is no reason to believe that the particles will be close to a Maxwellian.

2.1.1.1 Properties of the Maxwellian Distribution Function

The local Maxwellian has the following properties:

$$n(\vec{x}, t) = n(\vec{x}, t)\langle 1\rangle, \qquad (2.15)$$

$$\vec{u}(\vec{x}, t) = \langle \vec{v}\rangle, \qquad (2.16)$$

$$\frac{3kT}{2} = \left\langle \frac{m_p v^2}{2}\right\rangle. \qquad (2.17)$$

For a gas or plasma with zero mean velocity ($\vec{u} = 0$), the distribution function can be broken up into three independent components,

$$f_M = g_M(v_{x_1})g_M(v_{x_2})g_M(v_{x_3}), \tag{2.18}$$

where the reduced distribution function g_M is given by

$$g_M(v_{x_i}) = n(\vec{x}, t) \left[\frac{m_p}{2\pi k T(\vec{x}, t)} \right]^{1/2} \exp\left[-m_p v_{x_i}^2 / 2kT(\vec{x}, t)\right]. \tag{2.19}$$

This factorization of f_M shows that the three velocity components are distributed independently and that each of the components has a Gaussian distribution. (*Note:* Space plasmas can indeed have different "temperatures" along preferred directions in a magnetic field.) The Gaussian has zero mean and dispersion given by

$$\langle v_{x_i}^2 \rangle = \frac{1}{n} \int_{-\infty}^{\infty} g(v_{x_i}) v_{x_i}^2 dv_{x_i} = \frac{kT}{m_p}. \tag{2.20}$$

The root-mean-square velocity is defined by

$$\langle v_{x_i}^2 \rangle^{1/2} = (kT/m_p)^{1/2}, \tag{2.21}$$

and the thermal velocity is defined by

$$v_{\text{th}} = (2kT/m_p)^{1/2}. \tag{2.22}$$

The root-mean-square velocity is illustrated in Figure 2.1.

Since the Maxwellian is dependent only on the magnitude of the velocity, it is useful to express it in spherical polar coordinates in velocity space. These are (v, θ, ϕ), where in terms of Cartesian coordinates $v^2 = v_{x_1}^2 + v_{x_2}^2 + v_{x_3}^2$, $v_{x_3} = v \cos\theta$, $v_{x_1} = v \sin\theta \cos\phi$, and $v_{x_2} = v \sin\theta \sin\phi$. The element of volume in velocity space is

$$dv_{x_1} dv_{x_2} d_{x_3} = v^2 \sin\theta \, d\theta \, d\phi \, dv, \tag{2.23}$$

and since f_M depends only on the speed v, it is possible to define the reduced distribution function F_M by

$$F_M(v)dv = \int_{-\pi}^{\pi} d\phi \int_0^{\pi} d\theta f_M(v) v^2 \sin\theta \, d\theta \, d\phi \, dv, \tag{2.24}$$

so that, using Eqs. (2.14) and (2.22), the speed distribution $F_M(v)$ is given by

$$F_M(v) = 4\pi \left(\frac{1}{\pi}\right)^{3/2} (v^2/v_{\text{th}}^2) \exp(-v^2/v_{\text{th}}^2)/v_{\text{th}}. \tag{2.25}$$

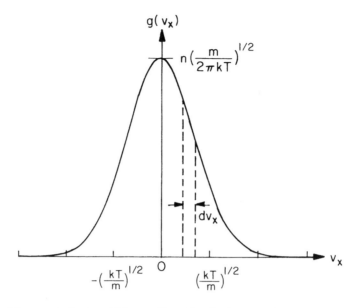

Figure 2.1. Reduced Maxwellian Distribution Relative to Velocity.

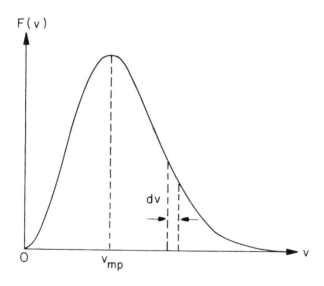

Figure 2.2. Speed Distribution Function.

F_M is plotted in Figure 2.2. The mean velocity $\langle v \rangle$ is

$$\langle v \rangle = 4\pi \left(\frac{1}{\pi}\right)^{3/2} \int_0^\infty v^2 / v_{\text{th}}^2 \exp(-v^2/v_{\text{th}}^2) \, dv / v_{\text{th}} = \left(\frac{8kT}{\pi m_p}\right)^{1/2}. \qquad (2.26)$$

This particular value of the mean velocity is often called $\bar{c}$. The most probable speed v_{mp} is determined from

$$\left. \frac{dF_M(v)}{dv} \right|_{v=v_{mp}} = 0.$$

This is easily shown to give $v_{mp} = v_{th}$. From this derivation, it can be seen that for any particle in a gas or plasma, the most likely speed it will have is v_{th}, whereas the average speed of all of the particles is $\bar{c}$.

For a gas or plasma with no directed motion $\vec{u} = 0$, it is obvious that $\langle v_{x_i} \rangle = 0$. However, it is of interest to consider the one-sided flux in a Maxwellian gas or plasma. This is the number of particles crossing a plane per unit area per unit time. Such a situation arises for the flux of particles to a spacecraft surface. The one-sided flux to a surface defined by the unit normal $\vec{n}$ is given by

$$\Gamma_n = \int_{\vec{v} \cdot \vec{n} > 0} f \vec{v} \cdot \vec{n} \, dv_{x_1} dv_{x_2} dv_{x_3}. \tag{2.27}$$

If the direction of v_{x_3} is taken to coincide with $\vec{n}$, then $\vec{v} \cdot \vec{n} = v \cos \theta$. Expressing the velocity space element in spherical polars gives

$$\Gamma_n = \int_0^\infty f_M v^3 dv \int_0^{\pi/2} \sin \theta \cos \theta d\theta \int_0^{2\pi} d\phi. \tag{2.28}$$

This integrates to

$$\Gamma_n = n \langle v \rangle / 4 = n\bar{c}/4. \tag{2.29}$$

The factor of $1/4$ can be understood as $1/2 \times 1/2$. The first $1/2$ arises from the choice of only those particles that are crossing the plane defined by $\vec{n}$, and the second $1/2$ arises from the average over the cosine distribution of velocities in the direction $\vec{n}$.

If a gas or plasma is in a conservative force field defined by

$$\vec{F} = -\nabla \Psi(\vec{x}), \tag{2.30}$$

then it is possible to analytically define the dependence of the density on the position vector. This is done from a consideration of momentum balance for the system. The equation for steady-state momentum balance is

$$0 = n\vec{F} - \nabla p, \tag{2.31}$$

where the pressure p is defined in Eq. (2.10). If the temperature is constant, then Eq. (2.31) can be integrated to give

$$n(\vec{x}) = n_0 \exp[-\Psi(\vec{x})/kT], \tag{2.32}$$

and n_0 is the number density in the region where $\Psi(\vec{x}) = 0$. The exponential in Eq. (2.32) is often called the Boltzmann factor, and the density is then said to be Boltzmann distributed. In this case, the local Maxwellian distribution function can be written as

$$f_M(\vec{x}, \vec{v}, t) = n_0 \left(\frac{m_p}{2\pi kT} \right)^{3/2} \exp\{-[m_p|\vec{v}|^2 + \Psi(\vec{x})]/2kT\}.$$

For a plasma in an electrostatic field, $\Psi = q\Phi$, where Φ is the electrostatic potential and q is the charge on the particles. This equation is an important relationship in computing spacecraft potentials in later chapters.

2.2 Typical Spacecraft Length and Velocity Scales

In order to analyze the physics of the interaction of the neutrals, plasma, radiation, or particulates with any spacecraft, it is important to define the typical length, velocity, and time scales associated with the perturbing body – the spacecraft. A comparison of the scales associated with a particular physical effect to the spacecraft scales will indicate the relative importance of each physical effect. For example, if the mean velocity with which the ambient neutral gas will expand into the rear of a rapidly moving spacecraft is v_{th}, and this is significantly smaller than the spacecraft velocity in orbit V_0, then little gas will get behind the spacecraft and the region will be relatively empty.

The typical size range of most spacecraft is from meters to many tens of meters. A typical GEO satellite is a few meters in height and radius, whereas the *Shuttle* is approximately 30 meters long and several meters across. The space station likely will be on the order of 100 meters long. With these dimensions, the body scale L_b can defined as $L_b \approx 1-100$ meters. (*Note:* The space tether or a long, thin boom may have scale sizes of a centimeter, but these are not discussed here.)

In any orbit, a spacecraft must move at the appropriate orbital velocity to stay up. The orbital velocity of a spacecraft on a circular orbit around the Earth is

$$V_0 = V_{cir} = \sqrt{GM_E/r},$$

where G is the gravitational constant, M_E is the mass of the Earth, and r is the radial distance from the center of the Earth. For LEO ($r \approx 6{,}600$ km), $V_0 \approx 8$ km/s; for GEO ($r \approx 42{,}000$ km), $V_0 \approx 3$ km/s. The escape velocity of a spacecraft from the Earth is

$$V_0 = \sqrt{2GM_E/R_E},$$

where R_E is the radius of the Earth. This gives an escape velocity of $V_0 = 11$ km/s. Thus, typical spacecraft velocities fall in the range $V_0 \approx 1-11$ km/s. With these length and velocity scales, the transit timescale and transit frequency can be defined. The transit time is

$$T_t = L_b/V_0, \tag{2.33}$$

and the transit frequency is

$$\nu_t = V_0/L_b. \tag{2.34}$$

On the basis of the numbers above, $T_t \approx 10^{-4}$–10^{-1} s and $\nu_t \approx 10$–10^4 Hz. The transit time is the timescale for the convective flow of the space gas or plasma past the spacecraft. It gives a measure of a timescale on which an interaction occurring at rest with respect to the Earth will be seen to move away from the spacecraft.

2.3 Neutral Gas Scales

2.3.1 Collision Mean Free Path and Knudsen Number

Neutral particles can only interact with themselves and a spacecraft through collisions. Therefore, to compare the effects of different neutral interactions, it is important to estimate the mean effect of collisions on the neutral gas. This is done through the concept of the mean free path. This concept is fundamental to the field of gas dynamics (Vincenti and Kruger, 1965; Bird, 1976) on which the presentation in this section is based.

In gas dynamics, the mean free path is defined as the average distance that a gas molecule travels between collisions. To define it mathematically, the concept of the collision cross section must be introduced. If two molecules of radii r_1 (called the test molecule) and r_2 (called the field molecule) are modeled as hard spheres, then the two molecules will collide when the distance between the centers is $r_1 + r_2$. If a tube is defined that has radius $r_1 + r_2$ with the field molecule centered in the tube, then any test molecule that enters the end of the tube will collide with the field molecule. The collision cross section for hard spheres is then defined as the cross-sectional area of the tube and is therefore

$$\sigma = \pi(r_1 + r_2)^2. \tag{2.35}$$

More generally, two molecules will interact whenever they approach each other sufficiently close that the strength of the intermolecular force is sufficient to modify the trajectories of the two particles. The typical intermolecular force field is plotted in Figure 2.3. The collision cross section is then defined as the cross-sectional area over which they interact. It is obvious from this definition and from Figure 2.3 that the cross-sectional area will, in general, be a function of the relative speed between the collision partners since particles that approach each other at a high

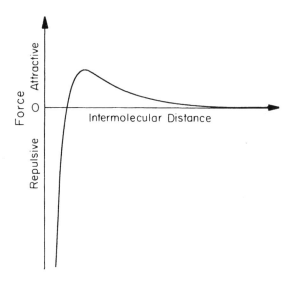

Figure 2.3. Intermolecular Force Field Relative to Distance.

relative velocity will be able to get closer to each other than slower particles before
they are repelled. In addition, it is clear that cross sections for different types of
interaction can be defined. A real molecule in a collision may collide elastically. It
also may be excited internally, dissociated chemically, undergo a charge exchange,
or be ionized. Each of these interactions has a collision cross section associated
with it, and a mean free path can be defined for each type of collision.

   With a definition of the collision cross section, the mean collision rate can be
defined. If a test molecule is injected into a distribution of field molecules, then the
test molecule will randomly collide with the field molecules. Take the number of
field molecules with velocities between $\vec{v}$ and $\vec{v} + \Delta \vec{v}$ to be Δn. If the test molecule
has velocity $\vec{v}_t$, then in a time interval Δt the test molecule will collide with any
molecule in a cylinder of volume $\sigma |\vec{v}_t - \vec{v}| \Delta t$. The number of field molecules in this
cylinder is Δn, so that the number of collisions in Δt is $\sigma |\vec{v}_t - \vec{v}| \Delta t \, \Delta n$. The mean
collision rate is then determined by averaging over all relative velocities and is

$$v = n\langle \sigma(g)g\rangle, \tag{2.36}$$

where g is the relative velocity $|\vec{v}_t - \vec{v}|$ and the average is defined in Eq. (2.7). The
mean free path is the average distance traveled by the test molecule between colli-
sions. The test molecule will have a mean speed $\langle |\vec{v}_t - \vec{u}| \rangle$ so that the mean free path is

$$\lambda_{\mathrm{mfp}} = \langle |\vec{v}_t - \vec{u}| \rangle / n\langle \sigma(g)g\rangle. \tag{2.37}$$

More generally, if molecules of type a are undergoing an interaction of class ξ (ξ
could be elastic collision, excitation, ionization, and so forth) with molecules of

type b, then the mean free path for type a is

$$\lambda_\xi^{ab} = \langle |\vec{v}_a - \vec{u}_a| \rangle / [n_b \langle \sigma_{ab}^{(\xi)}(g_{ab}) g_{ab} \rangle], \tag{2.38}$$

where $g_{ab} = |\vec{v}_a - \vec{v}_b|$.

The Knudsen number, K_n, in a gas flow is then defined as

$$K_n = \lambda_{\mathrm{mfp}}/L_b, \tag{2.39}$$

where the interaction is taken as elastic collisions. If the Knudsen number is much smaller than one, then a gas molecule suffers many collisions in a body scale length. In that case, the gas interaction with the body is well described by the continuum Navier–Stokes equations (Vincenti and Kruger, 1965). The usual criterion (Bird, 1976) for the validity of the continuum approach is taken as $K_n \leq 0.1$. For $K_n \leq 0.1$, the flow is defined as continuum or sometimes as collisional. For $K_n \gg 1$, the flow over a body is defined as collisionless and the individual molecules essentially follow ballistic trajectories determined by their inertia on the body scale. Flows with $K_n \approx O(1)$ are called transitional flows where both ballistic motion and a few collisions may be important to analyzing the flows. For neutral gas interactions with spacecraft, the flows are either collisionless or occasionally transitional.

2.3.2 Speed Ratio

For a neutral gas with a Maxwellian distribution and mean velocity $\vec{u}$, it is important to define the speed ratio between the mean speed and the thermal velocity. This is defined as

$$S = |\vec{u}|/v_{\mathrm{th}}. \tag{2.40}$$

If the gas has $S \gg 1$, then the motion of the gas molecules is dominated by the directed velocity $\vec{u}$ with a small random thermal component. This is the case in the flow of the ambient atmosphere ($|\vec{u}| = V_0$) past a LEO spacecraft. The spacecraft will encounter a strong flux on the side facing into the velocity vector (the ram side) and will experience very low fluxes on the side facing away from the velocity vector (the wake side). In the limit $S \ll 1$, the motion of the gas molecules is dominated by the thermal component and has only a small directed component. In this case, there will be little difference between the ram and the wake surfaces of the spacecraft.

It is of interest to note that the definition for S is almost the same as the definition for the Mach number in a continuum gas [$S = u/(2kT/m_p)^{1/2}$ and $M = u/(\gamma kT/m_p)^{1/2}$, where γ is the ratio of the specific heats in a gas and is of $O(1)$]. For the collisionless flows most often found in space, the concept of a speed of sound in the gas cannot be defined meaningfully on the scale L_b; therefore, the

Mach number is not a useful parameter. Nevertheless, the two definitions are sufficiently similar that analogies can be drawn between them. In particular, the physics of the flow as described above for $S \gg 1$ and for $S \ll 1$ correspond to the well-known behavior for supersonic and subsonic flows in a continuum gas. Therefore, the case $S \gg 1$ is sometimes referred to as supersonic flow, and the case $S \ll 1$ is likewise referred to as subsonic flow.

2.4 Plasma Scales

2.4.1 Basic Particle Motion in Constant Electric and Magnetic Fields

In contrast to neutral particles, which move under the influence of their inertia and collisions, charged particles also move under the influence of electromagnetic forces. Except near the upper fringes of the Earth's atmosphere, collisional and frictional forces on particles can, in general, be ignored. The two main forces on particles, F, then, are the electrostatic force,

$$\vec{F}_E = q\vec{E}, \tag{2.41}$$

and the magnetic (Lorentz) force (MKS units),

$$\vec{F}_B = q(\vec{v} \times \vec{B}), \tag{2.42}$$

where q is the particle charge (including sign, $q = Ze$ where e is the absolute value of the charge on an electron equal to 1.60219×10^{-19} Coulomb, and Z is the charge number equal to -1 for an electron, $+1$ for a singly charged ion, $+2$ for a doubly charged ion, and so forth), $\vec{v}$ is the velocity vector of the particle, $\vec{B}$ is the magnetic-field vector in space with magnitude B, and $\vec{E}$ is the electric-field vector in space with magnitude E.

Consider the actual motion of a particle subject to the forces in Eqs. (2.41) and (2.42). When the electric field is set to zero and the definition of the cross product is used, Eq. (2.42) implies that the force on a charged particle is always perpendicular to both the particle's instantaneous-velocity vector and the magnetic-field vector. This means that a particle must, in the absence of another force and in the presence of a uniform magnetic field, move in a circle in the plane perpendicular to the magnetic-field vector. It may additionally move freely (without any acceleration) along the magnetic field, mapping out a helix around its center of motion [see Bittencourt (1986)]. The radius ρ_p (called the cyclotron or gyro radius) of this circle is found by equating the centripetal force, $m_p v_\perp^2/\rho_p$, to the Lorentz force. In this expression, m_p is the particle mass and $v_\perp$ is the component of the velocity perpendicular to B. The expression is:

$$\rho_p = \frac{m_p v_\perp}{qB}. \tag{2.43}$$

The frequency with which the charged particle gyrates – the cyclotron frequency, Ω_p – is given by (MKS units)

$$\Omega_p = q B / m_p, \qquad (2.44)$$

where Ω_p is in radians per s.

According to Eq. (2.42), any particle motion parallel to B is unaffected by B. The particle's motion can be described in terms of a velocity parallel to the field, $v_\parallel$, a velocity perpendicular to the field, $v_\perp$, and a quantity called the particle pitch angle, α, the angle that the particle motion makes relative to the B direction:

$$\alpha = \text{arc } \cos(v_\parallel / v). \qquad (2.45)$$

The motion of the particle can be pictured as spiraling along the magnetic-field direction and executing a cyclotron motion around the field while moving along the field. A charged particle will deviate from these simple motions if there is an electric field or if the magnetic field has temporal or spatial changes. As an example, consider the case where there is a constant electric field. In this case, the particle equation of motion is

$$m_p \frac{d\vec{v}}{dt} - q(\vec{v} \times \vec{B}) = q \vec{E}. \qquad (2.46)$$

The homogeneous solution to this equation of motion is the cyclotron motion discussed previously. A particular solution to this equation for constant electric field perpendicular to $\vec{B}$ is (MKS units)

$$\vec{v} = \frac{\vec{E}_\perp \times \vec{B}}{B^2} = \vec{v}_d. \qquad (2.47)$$

This velocity corresponds to the so-called drift velocity of the particle. The total velocity of the particle is then composed of a gyration about the magnetic field, unimpeded flow parallel to the magnetic field (for $\vec{E} \cdot \vec{B} = 0$), and a drift perpendicular to both the electric and the magnetic fields. This drift is known as the "E cross B" drift velocity. Because it is charge- and mass-independent, both electrons and ions have the same drift, and there is no net current.

The gyro radius of electrons in a plasma is smaller than the gyro radius of ions in a plasma by the mass ratio m_e / m_i. Therefore, the motion of the electrons in this constant electric- and magnetic-field configuration will be to perform a tight cycloidal motion across the field because of the combination of the cyclotron and "E cross B" motion. The ions will perform, as illustrated in Figure 2.4, a much larger cycloidal motion than the electrons.

A measure of the importance of magnetic effects in the plasma around a spacecraft can be obtained by defining the ion and electron magnetization parameters. For a

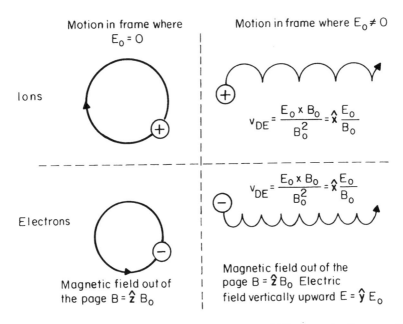

Figure 2.4. Cycloidal Trajectories Described by Ions and Electrons in Constant Electric and Magnetic Fields.

Maxwellian plasma, the electron magnetization parameter, M_e, is defined as

$$M_e = \rho_e / L_b = \frac{m_e v_{\text{th}_e}}{eBL_b}, \qquad (2.48)$$

which is the ratio of the gyro radius, defined for the thermal velocity, to the spacecraft dimensions. If $M_e \ll 1$, then the electron flow around the body is dominated by magnetic effects such as the cyclotron motion. The ion magnetization parameter is defined similarly as

$$M_i = \rho_i / L_b = \frac{m_i v_{\text{th}_i}}{qBL_b}. \qquad (2.49)$$

If $M_i \gg 1$, then the typical ion trajectories will be seen as ballistic on the scale of the spacecraft, and magnetic effects can be ignored for the ions. A plasma flow with both $M_e \ll 1$ and $M_i \ll 1$ is said to be magnetized, whereas $M_e \gg 1$ and $M_i \gg 1$ is said to be unmagnetized. Note that for $T_e \approx T_i$, $M_e/M_i \approx (m_e/m_i)^{1/2}$. Thus if $M_i \ll 1$, then $M_e \ll 1$. Since the magnetization parameters scale as $1/B$, in regions of low magnetic field the plasma flow typically will be unmagnetized.

2.4.2 Debye Length and Natural Plasma Frequencies

A plasma is defined as a collection of charged particles whose behavior is dominated by collective particle interactions (Bittencourt, 1986). On a large enough scale, a

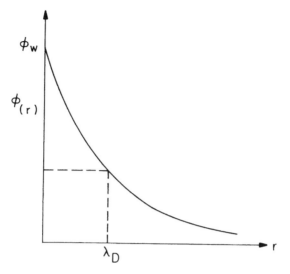

Figure 2.5. Debye Shielding in a Plasma.

plasma that is near equilibrium must be approximately charge neutral. If this were not the case, the strong Coulomb electrostatic interactions would drive the particles apart and not allow an equilibrium state to exist. The length scale over which this charge neutrality is established in a plasma is known as the Debye length. This characteristic length can be derived from the following simple argument.

Consider the surface of a spacecraft in a plasma and assume that the surface potential is Φ_w. This situation is illustrated in Figure 2.5. The potential in the plasma in front of the wall is given by the Poisson equation (Bittencourt, 1986)

$$\nabla^2 \Phi = e(n_e - n_i)/\epsilon_0, \tag{2.50}$$

where the ion plasma density is n_i, the electron plasma density is n_e, and ϵ_0 is the dielectric constant of free space. If the electrons and ions are distributed with a Maxwellian distribution in the plasma and have the same temperature kT, then the density of the electrons is determined from Eq. (2.32):

$$n_e = n_{e_0} \exp(-e\Phi/kT). \tag{2.51}$$

Similarly, the ion density is

$$n_i = n_{i_0} \exp(e\Phi/kT). \tag{2.52}$$

At infinity, the potential of space is defined as zero and it must be the case that the density of ions and electrons is equal there (if this were not the case, the electrostatic potential would not be zero). Therefore $n_{e_0} = n_{i_0} = n_0$. The use of these densities in

Eq. (2.50) then gives

$$\nabla^2\Phi = en_0[\exp(-e\Phi/kT) - \exp(e\Phi/kT)]/\epsilon_0. \qquad (2.53)$$

This equation can be normalized by defining $\tilde{\Phi} = e\Phi/kT$, the length λ_D by (MKS units)

$$\lambda_D = \sqrt{\epsilon_0 kT/(n_0 e^2)}, \qquad (2.54)$$

and $\tilde{x} = x/\lambda_D$. The equation then becomes

$$\frac{d^2\tilde{\Phi}}{d\tilde{x}^2} = [\exp(-\tilde{\Phi}) - \exp(\tilde{\Phi})]. \qquad (2.55)$$

This equation must be solved subject to the boundary conditions that at $\tilde{x} = 0$, $\tilde{\Phi} = e\Phi_w/kT$, and at infinity $\tilde{\Phi} = 0$. If the plasma has a potential energy small compared to its thermal energy ($e\Phi/kT \ll 1$), then a Taylor series expansion of the right-hand side of Eq. (2.55) gives

$$\frac{d^2\tilde{\Phi}}{d\tilde{x}^2} = -2\tilde{\Phi}. \qquad (2.56)$$

This gives the solution, which satisfies the boundary conditions of

$$\Phi = \Phi_w \exp[-x/(2\lambda_D)]. \qquad (2.57)$$

This potential is illustrated in Figure 2.5. The length λ_D is called the Debye length. Based on this derivation, the Debye length is the length scale over which the effect of the spacecraft surface potential is shielded in the plasma. The Debye length is also the length scale over which the plasma is approximately charge neutral. This transition zone gives rise to the important concept known as quasi-neutrality. That is, on length scales smaller than the Debye length, the plasma may not be charge neutral. This is manifestly true near the spacecraft surface, which will attract one species and repel the others. Mathematically, this can be expressed as

$$qn_i - en_e = 0 \qquad (2.58)$$

for length scales longer than a Debye length. The region near the spacecraft surface where the quasi-neutrality is violated is known as the sheath. It can be regarded as the analog in a plasma to a boundary layer in a classic fluid, and quasi-neutrality can be regarded as the outer solution of the flow. The Debye length is also the length scale over which collective plasma effects are manifested. For length scales smaller than the Debye length, the charged particles behave as individual charged particles subject to electrical effects. On longer length scales, collective coupling of particles can occur. The existence of this fundamental length scale allows the definition of an electrical coupling parameter for a spacecraft interacting with the space plasma.

A plasma–spacecraft interaction is said to be electrically coupled if $\lambda_D \gg L_b$ and electrically uncoupled if $\lambda_D \ll L_b$. If $\lambda_D \gg L_b$, a potential on one part of the spacecraft will be felt all over the spacecraft. In the opposite limit, each part of the spacecraft will only couple to the surfaces in its immediate vicinity.

In addition to electrostatic effects, there are numerous electromagnetic phenomena present in charged plasmas. Wave excitation and propagation are the central electromagnetic phenomena associated with plasmas. The basic plasma wave parameter is the electron plasma frequency, ω_{p_e}, given by (MKS units):

$$\omega_{p_e} = \sqrt{e^2 n_e / (\epsilon_0 m_e)}. \tag{2.59}$$

This is the fundamental oscillation frequency in a quasi-neutral plasma. It can be seen from the definition of the Debye length and electron thermal velocity that

$$\omega_{p_e}^2 = \frac{v_{th_e}^2}{2\lambda_D^2}. \tag{2.60}$$

Thus the electron plasma frequency can be understood as the frequency that the plasma will oscillate if a thermal fluctuation separates the ions from the electrons by a Debye length. The other fundamental frequency, the ion plasma frequency, is defined for a single ion species as

$$\omega_{p_i} = \sqrt{q^2 n_i / (\epsilon_0 m_i)}, \tag{2.61}$$

where the charge q is the total charge on the ions.

Of additional importance in any discussion of wave phenomena are the lower hybrid frequency, ω_{lh}, and the upper hybrid frequency, ω_{uh}. They are essentially mixtures of the plasma and cyclotron frequencies of the electrons and the ions, and are given by:

$$\omega_{lh} = \sqrt{\omega_{p_i}^2 / \left(1 + \omega_{p_e}^2 / \Omega_e^2\right)} \tag{2.62}$$

and

$$\omega_{uh} = \sqrt{\omega_{p_e}^2 + \Omega_e^2} \tag{2.63}$$

The kinetic motions of plasma particles are characterized by the frequency of their collisions. Electrons in the space plasma can collide with other electrons, ions, and neutral particles. The complete expressions for the collisions of electrons with either other electrons or ions are very complex (Trubnikov, 1965) and depend on the relative energies of the ions and electrons as well as the charge. For a Maxwellian plasma where the ions and electrons have velocities such that the electron thermal velocity is the largest average velocity (e.g., $v_{th_i} \ll V_0 \ll v_{th_e}$) and the ions are singly charged, the electron-charged particle collision frequency is given by Chen (1984) as

$$\nu_e = \nu_{ei} + \nu_{ee} = 2.9 \times 10^{-12} n_e \ln \Lambda_D (kT_e/e)^{-3/2} \ s^{-1}, \qquad (2.64)$$

where ν_{ee} is the electron–electron collision frequency, ν_{ei} is the electron–ion collision frequency, n_e is the electron density per cubic meter, and T_e is the electron temperature in degrees Kelvin. The logarithm $\ln \Lambda_D$ is called the Coulomb logarithm and is the result of the cutoff in a plasma of the particle interactions because of Debye shielding. The logarithm is only weakly dependent on the plasma parameters and typically ranges from 10 to 20.

In a partially ionized gas such as that in LEO, the electrons also can collide with the neutrals. The electron–neutral collision frequency, ν_{en}, for a Maxwellian plasma is given by

$$\nu_{en} = n_n \sigma_{en} v_{th_e} / \sqrt{2} \ s^{-1}, \qquad (2.65)$$

where n_n is the neutral density and σ_{en} is the electron–neutral collision cross section. This has been measured for a large number of gas molecules and is typically of $O(5 \times 10^{-19})$ m² although it can be energy dependent. In that case, the cross section must be averaged over energy. In a partially ionized gas, the total electron collision frequency is $\nu_e + \nu_{en}$. The mean free path for electrons in such a gas and/or plasma mixture can then be defined as $\lambda_{mfp}^e = v_{th_e} / (\nu_e + \nu_{en})$.

The ion collision frequency in a Maxwellian plasma is given by

$$\nu_i = \nu_{ii} + \nu_{ie} = 4.8 \times 10^{-14} n_i \ln \Lambda_D (kT_i/e)^{-3/2} \mu^{-1/2} \ s^{-1}, \qquad (2.66)$$

where ν_{ii} is the ion–ion collision frequency, ν_{ie} is the ion–electron collision frequency, and the collision frequency is dominated by the ion–ion collisions. In Eq. (2.66), T_i is the ion temperature in degrees Kelvin, and $\mu = m_i / m_{proton}$ or the ion mass in atomic mass units. The ion–neutral collision frequency in a Maxwellian gas and/or plasma mixture is

$$\nu_{in} = n_n \sigma_{in} v_{th_i} / \sqrt{2} \ s^{-1}, \qquad (2.67)$$

where σ_{in} is the ion–neutral collision cross section. This also has been measured for a large number of gas molecules and is typically of $O(5 \times 10^{-19})$ m². Similar to the electrons, the mean free path for ion collisions in a partially ionized gas is $\lambda_{mfp}^i = v_{th_i} / (\nu_i + \nu_{in})$.

A plasma–spacecraft interaction is said to be collisionless if, for both ions and electrons, $\lambda_{mfp} \gg L_b$ and collisional if the opposite is true. For a collisionless interaction on the scale of the body, the ions and electrons will mainly be streaming past the body, and collisions will have only a small effect on the plasma interactions. For the parameters found in space, the plasma–spacecraft interactions are usually considered collisionless for practical purposes.

2.4.3 Speed Ratios

In a neutral gas, there is only one possible thermal velocity and, therefore, only one possible definition for the speed ratio. The situation is more complex in a plasma in that two speed ratios can be defined. These are the electron speed ratio, S_e, given by $S_e = |\vec{u}_e|/v_{\text{th}_e}$, and the ion speed ratio defined by $S_i = |\vec{u}_i|/v_{\text{th}_i}$. These two speed ratios allow the definition of a new interaction regime. If the plasma flow around a spacecraft satisfies $S_i \gg 1$ and $S_e \ll 1$, then the flow is said to be mesothermal or mesosonic. In this case, the ions behave supersonically and the electrons subsonically. The other two regimes are $S_i \ll 1$ and $S_e \ll 1$ (subsonic) and $S_i \gg 1$ and $S_e \gg 1$ (supersonic). In subsequent chapters, the characteristics of plasma interactions for these regimes are compared and contrasted.

2.5 Radiation Invariants

2.5.1 Adiabatic Invariants

In Section 2.4.1, it was shown that in a constant electric and magnetic field, the motion of a charged particle is a gyration around the magnetic field, unhindered flow along the magnetic field, and drift perpendicular to the magnetic and electric fields with a constant drift velocity. Real electric and magnetic fields associated with the planets are not constant, but slowly vary on time scales such that the change of the field over a cyclotron orbit is small

$$\frac{1}{\Omega_c}\left|\frac{\partial \ln B}{\partial t}\right| \ll 1.$$

For electric and magnetic fields that vary on time scales less than certain characteristic times, charged particles in those fields have been shown to move in such a way that their motion will conserve three quantities (Krall and Trivelpiece, 1973; Jursa, 1985; Bittencourt, 1986). This conservation forms the foundation of the notion of adiabatic invariance and means that on some timescale there is a physical quantity that does not change significantly on that timescale.

The first adiabatic invariant is the magnetic moment, μ, which is conserved for variations on a timescale long compared with that of cyclotron motion. The magnetic moment is defined as

$$\mu = \frac{p_{\perp}^2}{2m_0 B}, \tag{2.68}$$

where p is the particle momentum, $p_{\perp}$ is the particle momentum perpendicular to the magnetic field, and m_0 is the particle rest mass. For nonrelativistic particles,

$$\mu = \frac{mv_{\perp}^2}{2B}, \tag{2.69}$$

whereas for relativistic particles,

$$\mu = \gamma^2 \frac{m_0 v_\perp^2}{2B},$$ (2.70)

where γ is the relativistic factor given by $\gamma = 1/(1 - v^2/c^2)^{1/2}$ and c is the speed of light.

In a magnetic field that increases in strength along the dominant magnetic-field direction, the invariance of the magnetic moment can be used to show that particles spiraling along the field will reflect at some point. This can be demonstrated by the following argument. The mirror ratio in a magnetic field, R_m, is defined as the ratio of the maximum field strength at which a particle reflects to the minimum field. If at the minimum value of the magnetic field, a particle has pitch angle α_0, then, from conservation of energy, the parallel velocity at any point along the field is

$$\frac{v_\parallel^2}{2} = \epsilon \cos^2 \alpha_0 + \mu B_{\min} - \mu B,$$ (2.71)

where ϵ is the particle energy per unit mass. Substituting the definitions of μ and R_m, this can be written

$$\frac{v_\parallel^2}{2} = \epsilon(1 - R_m \sin^2 \alpha_0).$$ (2.72)

Since $v_\parallel^2$ can never be negative, this implies that particles with a pitch angle at the minimum field greater than $(R_m)^{1/2}$ can never reach the point of maximum field. Therefore, they must reflect and reverse direction when $v_\parallel = 0$, which is the point where the field strength is given by

$$B = B_m = B_{\min}/\sin^2 \alpha_0.$$ (2.73)

This is known as the mirror equation. The argument demonstrates that in a magnetic field of increasing strength, the particle motion will be to gyrate around the field and bounce between regions of high magnetic field. The bounce time between two turning points is

$$\tau_B = \int_{-l_m}^{l_m} \frac{dl}{v_\parallel},$$

where dl is the element of length along a magnetic-field line and l_m is the distance from the minimum field to the field point where the particle mirrors. If the bounce time is long compared to the cyclotron period but short compared to temporal

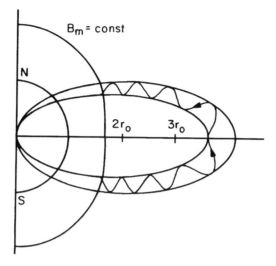

Figure 2.6. Motion of a Charged Particle in a Dipole Magnetic Field.

changes in the mirror locations, then it can be shown that there is a second adiabatic invariant called the longitudinal invariant, J. It is defined as

$$J = \int_{-l_m}^{l_m} p_{\parallel} \, dl. \tag{2.74}$$

Surfaces of constant J define the flux tube that bounds the cyclotron and the bounce motion of a given particle. This is illustrated in Figure 2.6 for a dipole field. Particles trapped on a given magnetic-field line in the dipole field will stay trapped on the field line despite slow temporal or spatial variations if the scale of the variations is slower than the bounce period.

2.5.1.1 B-L Coordinates and the Concept of Rigidity

It is shown in the next chapter that, near the Earth, the geomagnetic field can be approximated by a tilted magnetic dipole. In spherical polar coordinates, r, θ_m, and ϕ_m, the three components of the magnetic field are

$$B_r = -\frac{M}{r^3} 2 \cos \theta_m,$$

$$B_{\theta_m} = -\frac{M}{r^3} \sin \theta_m, \tag{2.75}$$

$$B_{\phi_m} = 0,$$

where M is the dipole moment given by $M = 0.311G \cdot R_E^3$ where R_E is the radius of the Earth ($R_E \approx 6400$ km). The equation for a magnetic-field line is given by

$$\frac{dr}{B_r} = \frac{rd\theta_m}{B_{\theta_m}} \tag{2.76}$$

and ϕ_m is a constant. This gives; for colatitude θ_m or latitude λ_m:

$$r = r_0 \sin^2 \theta_m = r_0 \cos^2 \lambda_m. \tag{2.77}$$

The distance r_0 is the radial distance at which a field line crosses the magnetic equator ($\theta_m = 90°$). The dimensionless distance, L, defined by

$$L = r_0/R_E, \tag{2.78}$$

is known as the McIlwain "L" parameter. The value of L is, by the definition of r_0, the equatorial crossing point of a magnetic-field line in terms of R_E. The L parameter can be used to define a surface called an L shell which, for a given L, is composed of all of the field lines that cross the magnetic equator at the given value of L. In terms of the L parameter, the dipole magnetic-field strength can be written as

$$B = \frac{B_E}{L^3} \frac{\sqrt{4 - 3\cos^2 \lambda_m}}{\cos^6 \lambda_m}, \tag{2.79}$$

where B_E is the magnetic-field strength on the Earth's surface at the magnetic equator.

In this magnetic field, the curvature of the field lines and the radial gradient in the field mean that the particles will feel forces normal to the magnetic-field direction as they gyrate and bounce between mirror points. The effect of these forces, like the force of an electric field normal to the magnetic field, will be to induce a particle drift motion perpendicular to the force and to the field. In the spherical coordinate system above, this drift will be in the azimuthal direction (ϕ_m). The azimuthal drift time around the Earth is given by

$$\tau_d = \oint ds/v_d,$$

where ds is measured around the drift orbit and v_d is the drift velocity. If the drift time, τ_d, is long compared with the bounce time, τ_B, but short compared with changes the overall field, then the third adiabatic invariant, the flux invariant, Φ, is defined by

$$\Phi = \int_S \vec{B} \cdot d\vec{S}, \tag{2.80}$$

where the surface S is the surface bounded by the azimuthal drift path. In the dipole field defined in Eq. (2.75), the flux invariant can be written as

$$\Phi = -\frac{2\pi B_E R_E^2}{L}. \tag{2.81}$$

The constancy of the flux invariant implies that the particles drift around the Earth on surfaces of constant L, which are the "L shells" defined earlier.

The motion of the particles in the Earth's field is thus characterized as gyration around the field line, bouncing between points given by B_m (which is constant as a consequence of the invariance of μ), and drifting around the Earth on constant L shells. The complete particle motion is illustrated schematically in Figure 2.6 by rotating the arc traced by the center of motion of the gyrating particle around the axis of the dipole field. This natural motion of the particles suggests a coordinate system defined by surfaces of constant magnetic-field magnitude, B, and shells of constant L. The intersection of a surface of constant B and an L shell is a closed curve around the dipole magnetic field. The position on this curve can be specified in terms of the magnetic longitude. The resulting coordinate system is known as the McIlwain B-L coordinate system. Because the particles in a dipole field drift on shells of constant L and bounce between points of constant B, the particle population can be completely described in terms of the particle flux as a function of just two coordinates – the B and the L values.

The three invariants together allow the development of simple, time-averaged models of the trapped radiation particle fluxes. It should be remembered, however, that for actual fields, perturbations in the electric and magnetic fields will modify the particle motion so that the three invariants are violated. There are indeed perturbations in the magnetosphere that occur on timescales of the cyclotron, the bounce, and the drift periods that violate the three invariants and cause deviations from the simple particle motion described here. Such variations lead to the diffusion of the trapped particle populations (primarily the electrons) in pitch angle.

The above analysis of particle motion holds for particles trapped in the geomagnetic field. The fact that particles bounce and drift on L shells gives rise to the concept of the radiation belts. This concept is discussed in detail in Chapter 3. In addition to the motion of these trapped particles, it is also important to estimate the ability of high-energy charged particles to penetrate from deep space to spacecraft orbits near the Earth. As the external charged particles encounter the Earth's magnetic field, their trajectories are modified by the Lorentz force. The measure of the resistance of a charged particle to the Lorentz force deflecting the particle trajectory from a straight line is called the magnetic rigidity, P. The rigidity has units of momentum per unit charge and is defined as

$$\vec{P} = \frac{\vec{p}}{q},\qquad(2.82)$$

where q is the charge on the particle. The equation of motion for a particle entering the Earth's magnetic field can be integrated to give

$$\vec{P} = \vec{P}_0 + \int_0^t \vec{v} \times \vec{B} dt.$$

Hence the Earth's magnetic field essentially acts as a momentum analyzer for high-energy particles impinging on it. For any point on the Earth's surface or in the magnetosphere at a given radius, latitude, and longitude, a minimum rigidity ($P_{0_{min}}$) can be defined along a given direction. Particles entering the Earth's magnetic field with a lower rigidity than this minimum will never reach an observer at the location for the specified direction since the Earth's field will deflect them before they arrive. The minimum rigidity is calculated by numerical integration of a particle trajectory through a model of the geomagnetic field. In a subsequent chapter, this rigidity will be shown to be an important quantity in determining what high-energy cosmic-ray particles can reach a spacecraft.

2.5.2 Linear Energy Transfer Distance

This section provides an overview of the basic physical concepts necessary to understand radiation penetration into matter. In particular, the concepts of energy, flux, fluence, and dosage are introduced.

Consider first the concept of energy. In the case of nonrelativistic particles, the fundamental equation relating particle mass and velocity to energy is:

$$E = \frac{1}{2}mv^2.\qquad(2.83)$$

For photons (which have no rest mass), the equivalent equation is

$$E = h\nu,\qquad(2.84)$$

where h is Planck's constant and ν is the frequency of the light.

Closely coupled to the concept of energy is that of dose. Simply put, dose is the total energy accumulated, due to the incident radiation, in a given volume element of a specific material. It is typically given in units of rads or radiation-absorbed dose for a particular material (the material must be specified because energy absorption is dependent on the material). As an example, for silicon, 1 rad (Si) = 100 ergs/g

(Si). The corresponding unit for dose in the MKS system is called the gray (Gy), that is, 1 Gy = 1 joule/kg = 100 rad = 10^4 ergs/g. It needs to be emphasized that, for the same incident flux, different materials will be affected differently depending on the composition of the incoming radiation and the composition of the absorbing material. Other units, such as the roentgen (quantity of γ rays or X rays that deposit, by ionization and energy absorption, 83 ergs/g in dry air) or the rem (roentgen equivalent per man), also are often used but are not discussed here.

In addition to the energy and composition of a particle or photon, it is also necessary to describe how many of them there are. This is usually done in terms of intensity or flux and, when speaking in terms of a time interval, fluence. Confusion arises over the concepts of intensity/flux and fluence because there are many different ways to define these quantities. Here, we define the quantity "unidirectional differential intensity" $j(E, \theta, \phi, t)$ as follows: the flux (number of particles or photons per unit time) of a given energy E per unit energy interval dE in a unit solid angle ($d\Omega = 2\pi \cos\theta d\theta d\phi$) about the direction of observation (in the θ, ϕ direction), incident on unit of surface area (dA) perpendicular to the direction of observation.

Typical units are particles/m^2-s-sr-KeV for protons or electrons and particles/m^2-s-sr-MeV-μ for heavy ions (where μ is nucleon). The unidirectional integral intensity (or flux) is defined as the intensity of all particles with energy greater than or equal to a threshold energy E:

$$j_{\geq E} = \int_E^\infty j \, dE, \tag{2.85}$$

with units of particles/m^2-s-sr.

The omnidirectional flux, J is defined as

$$J = \int_{4\pi} j \, d\Omega. \tag{2.86}$$

The fluence, I, is the integral of the flux over a given time interval (e.g., 1 hour, 1 year):

$$I = \int_{\delta t} j \, dt. \tag{2.87}$$

Here, a reference to the omnidirectional fluence $I(>E)$, means the "omnidirectional integral (in energy) fluence" such that:

$$I_{>E} = \int_E^\infty dE \int_{4\pi} d\Omega \int_{\delta t} j \, dt. \tag{2.88}$$

The units of this quantity are particles/m^2 for some specified threshold energy E (typically 1 MeV or higher for radiation effects) and for a specified time interval (often 1 year).

To allow comparisons between different energies, particle types, and dosages, it is common practice to talk in terms of "1-MeV equivalent" (typically 1-MeV electrons or 1-MeV neutrons in silicon). The energy dependence of the damage and energy content of the spectra for the environment to be considered are used to determine what fluence of 1-MeV particles (electrons or neutrons) would produce the same amount of damage or dose in the material (typically silicon or aluminum). (*Note:* Because of variations in the damage parameter with material and property, the use of a damage equivalent is not exact but an approximation for comparison purposes.)

A final quantity related to energy absorption and flux is the linear energy transfer (LET). LET is the energy transferred by radiation per unit length of absorbing material. That is, LET $= dE/dx$. For ionization and excitation effects, it is often expressed in MeV/μm of the primary particle track length or, if the density of the material is known, MeV-cm^2/mg [this is typically the unit when the reference is to an LET between 1 and 30 and is given by $1/\rho(dE/dx)$].

The concept of LET is particularly important when discussing single-event upsets (SEU) or "soft errors." These occur when a particle, typically an ionized, high-energy atomic nucleus, deposits enough energy in the sensitive region of an electronic device to cause a change in the logic state of the device. Upsets occur only when the energy deposited exceeds a critical level in the sensitive region of the device. This is often computed in terms of LET. When viewed as a function of LET, the probability of upset is, in its simplest form, a threshold phenomenon: Any particle with a minimum LET (L_0) or greater will cause an upset. This behavior is illustrated in Figure 2.7 where the energy deposited per unit length (LET) is plotted

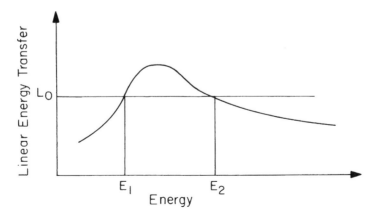

Figure 2.7. LET Relative to Energy.

versus incident particle energy (note that the curve has a peak rate). A useful way of presenting the environment in terms of LET is the Heinrich curve. The Heinrich curve gives the integral flux as a function of LET rather than particle energy. The Heinrich flux, F_H, is the flux of particles for a single species with a (threshold) LET of LET_0 or greater:

$$F_H(\text{LET}_0) = \int_{E_1}^{E_2} f(E)\,dE, \tag{2.89}$$

where f is the particle flux for the species as a function of energy, and E_1 and E_2 are the energies between which the LET is greater than or equal to the threshold LET_0. The LET depends not only on particle energy but the target material as well. Thus the LET-versus-energy curve will be different for all particle species. Experiments have shown, however, that, to first order, it is the LET that is important for determining upsets and not the particle energy or its species. The Heinrich flux-versus-LET plot is the principal means of presenting radiation data for use in SEU calculations just as the particle flux versus energy is the main means of presenting radiation data for dosage calculations.

3

The Ambient Space Environment

In this chapter, the principal natural (unperturbed) environments responsible for spacecraft interactions are introduced. These are the solar environment, the neutral atmosphere, the geomagnetic field, the plasma environment, the geostationary environment, energetic particle radiation, electromagnetic and optical radiation, and particulates (debris and meteoroids). The ambient space environment defined by these components has been the subject of numerous books and review papers [e.g., Jursa (1985)] or the excellent short descriptions of the environment in MIL-STD-1809 (1991). Unlike most of these sources, which deal primarily with the details of the space environment, the intent here is to provide the reader with sufficient background to evaluate the potential impact of the environment – both natural and man-made – on a spacecraft.

The relationships between the orbit classes and the natural environment are summarized in Figure 3.1. Table 3.1 is used to indicate which environments must be considered for a given class of orbits.

3.1 Influence of the Sun

The dominant energy source for the space environment in the solar system is the Sun. The chief solar influence on the space environment is through its electromagnetic flux (see Section 3.4.2) and the charged particles that it emits. The solar particle flux is composed basically of two components: the very sporadic, high-energy ($E > 1$ MeV) plasma bursts associated with solar events (flares, coronal mass ejections, proton events, and so forth) and the variable, low-energy ($E \approx$ tens of eV) background plasma referred to as the solar wind. Although the electromagnetic flux can penetrate to the Earth's atmosphere and, at many wavelengths, to the Earth's surface, the charged-particle environment is largely shielded by the Earth's magnetosphere. The solar wind, because of its density (tens of particles per cm³)

Table 3.1. *Relationship between ambient environments and orbits*

Ambient environment	Orbits
Neutral	LEO and PEO
Plasma	LEO, PEO, MEO, GEO, and interplanetary
Radiation	Somewhat for LEO; PEO, MEO, GEO, and interplanetary
Particulates	LEO, PEO, and, interplanetary; somewhat for MEO, GEO

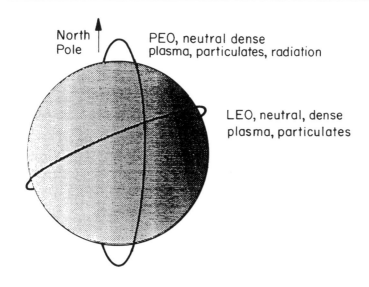

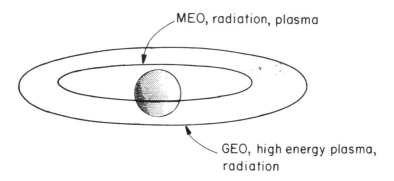

Figure 3.1. Schematic of Environments for LEO, PEO, MEO, and GEO.

and velocity ($\approx$200–2,000 km/s), energetically dominates the interplanetary environment and can directly reach the GEO environment on occasion. Although it does not contribute directly to the environment at LEO, PEO, or MEO, it does, as the primary energy source for geomagnetic activity, significantly influence these orbital environments.

3.1.1 Solar-Cycle Effects

Several hundred years of actual observations of the Sun have revealed a changing pattern of disturbances (although initially only sunspots were observable, starting in the early 1900s, flares and solar magnetic fields have been systematically recorded) that appear to follow semiregular patterns of about 11 years. These solar cycles, although fairly predictable in time, have peak activity levels that may vary by factors of four from one maximum to the next. Moreover, historical records indicate relatively long periods, for example, most of the seventeenth century (the so-called Maunder Minimum), during which there were no discernible solar cycles. Long-term variations in the neutral atmosphere and ionosphere of the Earth are observed in response to these cycles. The variations are apparently related to increases in the extreme ultraviolet (EUV) radiation flux from the Sun and increases in geomagnetic activity related to variations in the solar wind. Increases in solar activity raise temperatures in the upper atmosphere, causing it to expand outward and increase the density at a given altitude. Similarly, the ionospheric density and temperature respond to solar-cycle variations.

Solar activity is characterized by several indices. These include daily, monthly, and annual averages of sunspot number (typically called the R value), the 10.7-cm radio flux (believed to be representative of the EUV flux and referred to as $F_{10.7}$), and, at the Earth's surface, various geomagnetic activity indices (see Section 3.1.2). Models of the neutral atmosphere and the ionosphere, which typically employ one or more of these indicators as measures of the effects of solar activity, are used to predict the environments presented in this chapter.

3.1.2 Solar and Geomagnetic Indices

The oldest and most basic indicator of solar activity is the so-called sunspot number. Sunspots are localized regions of cooler solar atmosphere that appear dark against the hotter solar surface and are normally associated with intensification of the solar magnetic field. The main measure of sunspots, the Wolf or Zurich or relative daily sunspot number (R), is not, as its name would imply, an actual count of the visible sunspots on the Sun but rather an index for the degree of spottiness on the Sun. It is computed as $R = k(10g + s)$ where s is the number of individual spots, g is the

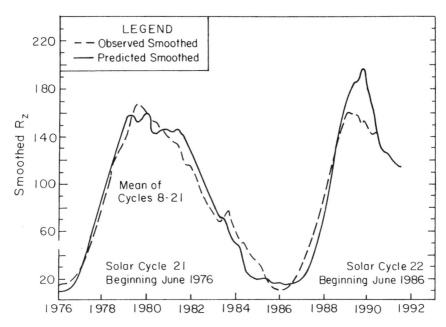

Figure 3.2. Observed and One-Year-Ahead Predicted Sunspot Numbers. From DeWitt, Dwight, and Hyder (1993). Reprinted by Permission of Kluwer Academic Publishers.

number of sunspot groups, and k is a subjective correction factor that allows for differences in equipment and observing conditions. The sunspot number is the most frequently used index for the general level of solar activity. From observations dating back to the eighteenth century, it has been found to fluctuate on an approximately 11-year cycle with maximum values of ≈ 150 and minimum values of ≈ 10. This behavior is illustrated in Figure 3.2.

Another important measure of solar activity is the $F_{10.7}$ flux. The $F_{10.7}$ index is the solar radio flux observed at a wavelength of 10.7 cm (2,800 MHz). It corresponds to a radio emission line for iron and is normally reported in solar flux units (sfu, but also called the jansky, j) where one solar flux unit is 10^{-22} W $\cdot$ m^{-2} $\cdot$ Hz^{-1}. Variations of $F_{10.7}$ are believed to follow variations in the solar EUV flux and to correlate with the overall long-term variations in solar activity. $F_{10.7}$ is used as a measure of solar activity effects on the atmosphere since it can be measured on the ground, whereas the EUV, which is believed to be the actual driver, cannot. It varies from about 50 sfu at solar minimum (R minimum) to 240 sfu at solar maximum (R maximum) and varies with the 11-year solar cycle.

Whereas sunspot number is used to represent the gross, long-term variation of solar activity, the short-term geomagnetic activity dependence is accounted for by the 3-hour, semilogarithmic K_p or its linearized form a_p. These indices represent magnetic-field disturbances induced by changes in the solar wind and, through

Table 3.2. *Variation in* A_p *and* a_p

Level	Quiet	Unsettled	Active	Minor storm	Major storm
Index	0–6	7–14	15–29	30–49	>50

heating effects, correlate with short-term variations of the upper atmosphere. The subscript p refers to planetary because the indices are the result of combining individual stations from around the world. Although the K_p index is the fundamental quantity, its linearized version, a_p, is more easily understood. The a_p values were selected so that they correspond to the maximum variations in the Earth's surface magnetic field at midlatitudes in a 3-hour period when the value is multiplied by 2γ ($1\gamma = 1$ nanotesla). The a_p and its daily average, A_p, range from a minimum value of zero to a maximum value of 400. Because K_p and a_p are ultimately related to solar activity, there is a weak correlation between them and R and $F_{10.7}$. The values of a_p and A_p can be qualitatively related to levels of geomagnetic activity by Table 3.2. Each value of a_p corresponds to one value of the K_p index. K_p is dimensionless and provides a semilogarithmic measure of the level of disturbance of the geomagnetic field during a 3-hour period. K_p ranges from 0_o to 9_o with 27 one-third unit steps ($0_o, 0_+, 1_-, 1_o, 1_+, 2_-, 2_o, 2_+$, and so forth).

As an example of the use of these indexes, the *space station* design requires that it be able to maintain attitude control against a worst-case neutral environment. This is defined in station documentation as $A_p = 140$ and $F_{10.7} = 230$.

3.1.3 Short-Term Events

Short-term variations in solar activity manifest themselves at LEO principally in terms of the effects of solar flares and geomagnetic storms. Solar flares (defined as an enhanced brightness in the solar light at the H_α frequency) sometimes, although not always, create energetic particle events. These energetic particles, coupled with changes in EUV flux that heat the atmosphere, occur far more frequently during solar maximum then solar minimum and last from a few minutes to a few hours.

The most dramatic solar-related changes in the Earth's environment at LEO are brought about by geomagnetic substorms. These changes are reflected in visible auroral displays and in intense particle and field variations in the auroral regions down to an altitude of 100 km. They are believed to result from a complex interaction between the solar wind plasma and magnetic field and the magnetosphere of the Earth. Variations in the solar wind, either in association with solar flares or changes in the solar-wind flow and magnetic-field direction, are the primary energy sources of these events. A correlation between the frequency of geomagnetic storms (as measured by a_p, K_p, and so forth) and the solar cycle exists but the intensity of these events is essentially independent of solar cycle. A typical geomagnetic storm often follows an abrupt change in the geomagnetic field (a storm sudden commencement or SSC) and can last several days. A geomagnetic storm is marked by a series of 0.5- to 2-hour impulsive events associated with auroral enhancements called substorms. These substorms correlate with the direction of the solar-wind magnetic field (or, equivalently, $\vec{V} \times \vec{B}$ electric field), the substorm occurring when the field turns southward.

3.2 The Neutral Atmosphere

In Figure 3.1, the environments are shown schematically for LEO and polar orbits. The ambient neutral atmosphere is an important environment at LEO and PEO altitudes. For the other orbits, the neutral density is too low to be of significance. Numerous types of models have been developed to describe the total density and/or composition and temperature of the upper atmosphere. At present, the most commonly used analytical model of the upper atmosphere is the Mass Spectrometer and Incoherent Scatter (MSIS) model (Hedin, 1983, 1987, 1991). This model, based on in-situ neutral mass spectrometry data, deviates on the average by about 20 percent from older models based on data from spacecraft drag. This is a relatively insignificant difference, given the large differences that can exist instantaneously between the models and the actual environment along a LEO spacecraft orbit (as much as a factor of 10 or more in some cases). In large part, this latter problem is due to the phase differences between when the models predict that geomagnetic storm-induced density changes will be observed at a given location and when they actually occur. To provide a quantitative estimate of the range of values to be expected, in Section 3.3, the neutral-atmosphere parameters for a nominal *space station* orbit and the *EOS* orbit (see Table 1.3) are estimated using the MSIS-86 model (1986 version).

Figures 3.3 and 3.4 show typical height profiles of temperature and composition, respectively, in the neutral atmosphere. The atmosphere can be divided into several characteristic regions, or spheres, based on the temperature or composition profiles or the physical processes operative in the region. Of particular relevance to LEO and polar operations is the thermosphere, which starts at an 85-km altitude and ends at 1,000 km. As shown in Figure 3.3, the thermosphere is characterized by a steady temperature increase with height – the actual structure is very dependent on the solar-activity level. In the thermosphere, particle collisions are insufficient to cause mixing of the species, and so hydrostatic equilibrium is achieved for each separate species. Above the thermosphere is the exosphere (particles in this

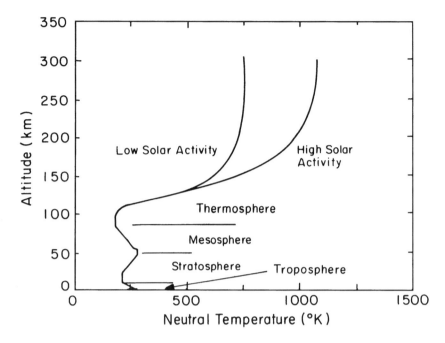

Figure 3.3. Temperature (K) Relative to Altitude (km) for the Neutral Atmosphere.

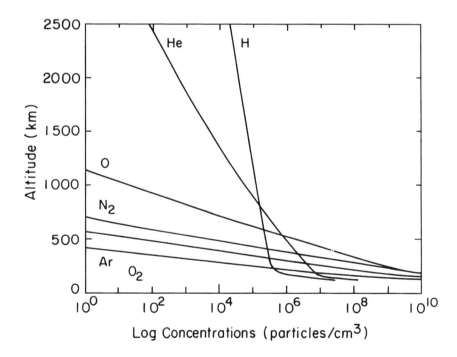

Figure 3.4. Composition (cm^{-3}) Relative to Altitude (km) for the Neutral Atmosphere.

region, primarily hydrogen and helium, are in individual orbits) where constant temperatures are expected.

Particle concentrations in the thermosphere decrease roughly exponentially with height as a result of the gases being in hydrostatic equilibrium. This is illustrated in Figure 3.4, where sample profiles of the major constituents during solar maximum and minimum are plotted. Ordered by decreasing mass (in amu), the major constituents of the neutral thermosphere are argon (Ar), 39; molecular oxygen (O_2), 32; molecular nitrogen (N_2), 28; atomic oxygen (O), 16; helium (He), 4; and atomic hydrogen (H), 1. Hydrostatic equilibrium makes the light particles, H and He, the dominant constituents at high altitude, whereas heavy particles, such as Ar, O_2, and N_2, are relatively unimportant. For most practical applications, the neutral atmosphere at LEO and polar altitudes is considered to consist mainly of atomic oxygen (*Note*: Atomic hydrogen can dominate occasionally above 500 km for low exospheric temperatures) with traces of molecular oxygen, molecular nitrogen, and atomic hydrogen. Helium, nitric oxide, atomic nitrogen, and argon are also present below the one percent level. The temperature of the constituents varies approximately exponentially from $\approx$100 K at 100 km to 500–1,500 K at 1,000 km depending on solar cycle, latitude, and local time with excursions to 2,000 K during high levels of geomagnetic activity. Since spacecraft between 100 and 1,000 km are moving at about 7.8 km/s, the resulting impact energy of the particles can reach values on the forward (or ram) surface of the spacecraft well in excess of 5 eV (varying from 4.6 eV for N to 10.25 eV for O_2). These ram energies are sufficiently high to induce chemical reactions (including oxygen erosion, as discussed in Section 4.4). Further, the large ratio of the directed velocity to thermal velocity means that pronounced anisotropies exist in the flux to the vehicle.

Values of the three basic neutral atmosphere parameters – concentration, temperature, and composition – vary in response to many factors, including local time, latitude, altitude, and solar and geomagnetic activity. For solar dependencies, the $F_{10.7}$ index usually represents the solar EUV radiation variations and the sunspot number, R, represents the solar influence on geomagnetic activity. EUV and geomagnetic-related heating are the dominant heat sources in the upper atmosphere. Observed values of neutral-atmosphere parameters, in general, increase with increasing $F_{10.7}$ and/or a_p/A_p indexes. For instance, the average neutral density at 400 km increases by about a factor of 10 from solar minimum ($F_{10.7} \approx 70$) to solar maximum ($F_{10.7} \approx 230$). At the same time, the exospheric temperature increases from $\approx$700 K to $\approx$1,200 K because of the extra EUV heating during solar maximum.

The time dependence of the neutral atmosphere reflects a seasonal pattern and a pronounced diurnal variation as the Earth rotates under the subsolar point. In addition, there is a semiannual pattern with the average total concentration at

Table 3.3. *Input conditions for MSIS-86 calculations*

	Max	Min	Mean
$F_{10.7}$	230	70	120
A_p	400	4	12
Month	Nov	July	May

Table 3.4. *MSIS-86 results for space station orbit*

Input Table 3.3	Output	n_n (m^{-3})	n_O (m^{-3})	T (K)	M (amu)
Max	MAX	5.5×10^{14}	5.3×10^{14}	2,023	17.4
	MIN	1.2×10^{14}	1.0×10^{14}	1,139	15.0
	AVE	3.2×10^{14}	2.8×10^{14}	1,535	16.2
Min	MAX	6.9×10^{12}	4.9×10^{12}	800	13.0
	MIN	1.3×10^{12}	0.5×10^{12}	626	7.3
	AVE	3.2×10^{12}	1.9×10^{12}	708	10.3
Mean	MAX	3.9×10^{13}	3.4×10^{13}	1,091	15.1
	MIN	0.9×10^{13}	0.7×10^{13}	847	13.1
	AVE	2.0×10^{13}	1.7×10^{13}	959	14.3

midthermospheric altitudes (500 km) at a maximum in October–November and in April. At the same altitudes, average daily temperatures maximize a few hours after noon, near 1400 hours local time.

In addition to standard inputs such as position and time, MSIS-86 model calculations require a substantial number of inputs including the solar $F_{10.7}$ and the geomagnetic (A_p) indexes. To illustrate the range of atmospheric changes induced by these variables, two sets of extreme values of input parameters along the *space station* and *EOS* orbits are considered here. A description of the orbits is given in Table 1.3. The corresponding input parameter values are listed in Table 3.3. The maximum solar and geomagnetic activity conditions are assumed to be $F_{10.7} = 230$ and $A_p = 400$. The input month is November (day 315) for the maximum and July (day 195) for the minimum conditions in the northern hemisphere for the seasonal variation, because these two months yield the highest and lowest particle concentrations, respectively. To allow a comparison between the input conditions, a mean set of parameters also is provided. The $F_{10.7}$ and A_p values assigned to the mean set are the averages over a solar cycle. The results for these input values are summarized in Tables 3.4 and 3.5 in terms of n_n (m^{-3}), the total concentration; n_O (m^{-3}), the atomic oxygen concentration; T (K), the thermospheric temperature; and M, the

Table 3.5. *MSIS-86 results for* EOS *orbit*

Input Table 3.3	Output	n_n (m^{-3})	n_O (m^{-3})	T (K)	M (amu)
Max	MAX	1.4×10^{14}	0.8×10^{14}	2,602	25.9
	MIN	1.0×10^{13}	0.7×10^{13}	1,206	11.3
	AVE	4.8×10^{13}	3.4×10^{13}	1,792	16.7
Min	MAX	2.5×10^{12}	0.1×10^{12}	893	9.1
	MIN	1.8×10^{11}	2.2×10^{9}	592	2.3
	AVE	8.1×10^{11}	0.3×10^{11}	726	4.1
Mean	MAX	4.0×10^{12}	2.2×10^{12}	1,228	14.4
	MIN	8.4×10^{11}	1.7×10^{11}	822	4.6
	AVE	2.2×10^{12}	7.5×10^{11}	995	8.5

mean atomic mass (amu). (*Note*: MAX and MIN are the maximum and minimum values computed for a quantity along the orbit, and AVE is the computed average for the orbit.)

From the results listed in Tables 3.4 and 3.5, note that both the total concentration and the atomic oxygen concentration are larger along the *space station* orbit than along the *EOS* orbit but the temperature has the opposite trend. In addition, the mean neutral mass shows a much larger fluctuation along the *EOS* orbit than for the *space station* orbit.

3.3 The Plasma Environment

In contrast to the ambient neutral environment, which is only significant for LEO and PEO, the space plasma environment can affect spacecraft in any orbit. Near the Earth, the plasma is cold and dense, but farther away, its density drops significantly. However, although its density drops, the mean energy of the plasma increases out to GEO, eventually transitioning to the solar-wind plasma as the magnetosphere is exited. Because the plasma is a collection of charged particles, it responds to magnetic-field variations. Thus it is typical to first describe the geomagnetic field which controls the plasma, then the plasma environment for LEO and PEO, then for MEO, and, finally, the GEO plasma environment (the solar-wind plasma characteristic of interplanetary orbits is described earlier). That is the approach followed here.

3.3.1 The Geomagnetic Field

Figure 3.5 illustrates schematically the Earth's plasma environment and the structure of the geomagnetic field for typical conditions. The geomagnetic field outside

EARTH'S MAGNETOSPHERE

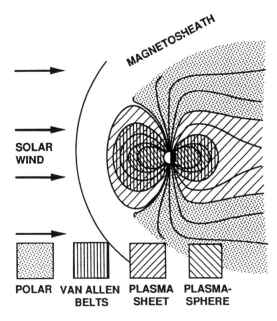

Figure 3.5. The Magnetosphere Showing the Plasmasphere, Plasma Sheet, and Polar Plasma Regions and the Van Allen Radiation Belts.

the plasmasphere deviates substantially from a simple dipole field. This is the result of the complex, magnetohydrodynamic interaction between Earth's magnetosphere and the solar wind. Above 1,000 km, the dominant geophysical environment is the magnetic field of the Earth – the source of the Earth's magnetosphere. Below 1,000 km, the Earth's magnetic field, primarily through the control of the ionospheric plasma, plays an important role in the dynamics of the natural environment, which tends to be dominated by the neutral atmosphere in this altitude regime. As such, an accurate knowledge of the ambient field is critical to a proper understanding of most plasma phenomena at all altitudes. The International Geomagnetic Reference Field (IGRF) is currently the principal model for determining the near-Earth magnetic field for altitudes up to 25,000 km. The following, based primarily on the IGRF model, describes the main characteristics of the near-Earth magnetic field.

The Earth's geomagnetic field, B, is composed of three distinct components (or current systems): the core field, the crustal field, and the external field (Jursa, 1985). The core field is the dominant field at the Earth's surface and is due primarily to the convective motion of the conducting fluid in the Earth's internal core. The crustal field, which originates in the region between the core–mantle interface and the

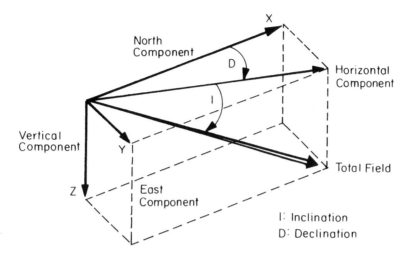

Figure 3.6. Geomagnetic-Field Magnetic Elements.

surface of the Earth, results from the remnant or induced magnetization of ferro-magnetic materials below their Curie temperature. The distribution of the crustal field results in surface anomalies that are associated with the geologic and tectonic features of the crust. For analytical purposes, the core and crustal fields are usually combined and referred to as the internal, or main, field, B_i. B_i varies slowly on the order of 100 years – currently ≈ 0.05 percent per year. B_i dominates at LEO and accounts for more than 99 percent of B even during extremely large geomagnetic storms. In contrast, the external field, B_e, which makes up ≈ 1 percent of the field at LEO altitudes, is due primarily to extraterrestrial sources – primarily, the ring current and the solar wind. It varies rapidly in time with variations from millisec-onds to 11 years (e.g., the solar cycle) and is closely correlated with geomagnetic activity and solar interactions.

The seven quantities, called magnetic elements, normally used to specify the geo-magnetic field (Jursa, 1985), are illustrated in Figure 3.6. As for any spatial vector, three independent quantities (e.g., $[H, D, Z]$ or $[X, Y, Z]$) are required to define B uniquely. For spacecraft operating in the LEO environment, the most convenient system to use is either the geographic (also known as geocentric) or the geomagnetic coordinate system. These systems, based on spherical coordinates, are schemati-cally defined in Figure 3.7. Geographic coordinates correspond to a geocentric longitude/latitude system based on the Greenwich prime meridian. Geomagnetic coordinates are similarly Earth centered, but the north pole of the system passes near the geomagnetic pole at latitude 78.5° and longitude 291.1° E. Geomagnetic longi-tude is measured from the great circle passing through the geographic and geomag-netic poles. Descriptions of other coordinate systems can be found in Jursa (1985).

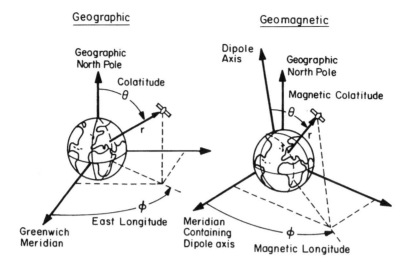

Figure 3.7. Spherical Coordinate System for Geomagnetic and Geographic Coordinates.

Aside from the gravitational field of the Earth, the magnetic field due to the internal geomagnetic field at LEO altitudes is the most accurately known of the natural environments. It can be modeled crudely in terms of a tilted ($-11°$ from geographic north) magnetic dipole of magnitude $M = 0.311G \cdot R_E^3 = 7.9 \times 10^{15}$ $\text{T} \cdot \text{m}^3$. In the geomagnetic coordinate system, the magnetic-field intensity induced by M at the point (r, θ_m, ϕ_m) is given by the expression:

$$B_i = (M/r^3)[3\cos^2(\theta_m) + 1]^{1/2} \tag{3.1}$$

and it has components given by

$$\begin{aligned} B_r &= -(M/r^3)2\cos\theta_m \\ B_{\theta_m} &= -(M/r^3)\sin\theta_m \\ B_{\phi_m} &= 0 \end{aligned} \tag{3.2}$$

B_i is found to have a maximum value of $\approx 0.6 \times 10^{-4}$ T near the polar cap and a minimum value of $\approx 0.3 \times 10^{-4}$ T near the equator at the Earth's surface.

Equation (3.1) is valid only for the idealized configuration of a centered dipole. Unfortunately, discrepancies as high as ± 25 percent exist between the measured field and the centered dipole field. Modifying the configuration to an eccentric dipole reduces the discrepancies to the 10 percent level. For most purposes, the much more accurate IGRF series of models is the internationally adopted standard. A version, IGRF-87, is a computer model based on a numerical fit of measured data to a magnetic scalar potential expanded in terms of 10 spherical harmonics (Barraclough, 1987). The model calculates the seven magnetic elements of B_i for

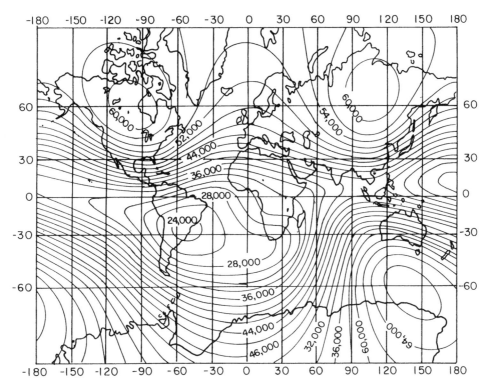

Figure 3.8. Total Intensity (F) in Nanoteslas. Reprinted from Jursa (1985) with Permission.

any given geographic location. Secular variations are included for input times up to the year 1995.

Figure 3.8 illustrates typical results from such calculations for the northern hemisphere at a constant altitude of 400 km. The field amplitude varies from a minimum of 0.25×10^{-4} T near the equator to 0.5×10^{-4} T over the polar caps. Two peaks exist in the magnitude of the magnetic field over the north pole (if vector components are considered, the maximum at 270° E is the true "dip" magnetic pole). Likewise, there are two minima near the equator – the largest of these is responsible for the so-called South Atlantic Anomaly, a region critical in determining radiation exposure for LEO. Geomagnetic storm variations are superimposed on this main field. These are typically less than 0.01×10^{-4} T, so that even during a severe geomagnetic storm, magnetic fluctuations are small at LEO altitudes compared to the average field (even though this is a very small change in the Earth's field, the effect of geomagnetic storms on particle fluxes in the polar ionosphere or at GEO can be tremendous).

For *space station* and *EOS* orbits, IGRF model calculations are summarized in Table 3.6. Along each orbit, values of the magnetic elements for a one-day mission

Table 3.6. *IGRF-87 results for* space station *and* EOS

	F (T)	X (North), T	Y (East), T	Z (Down), T
SS	0.23×10^{-4}	0.23×10^{-4}	-0.6×10^{-8}	-0.15×10^{-5}
EOS	0.14×10^{-4}	0.14×10^{-4}	0.3×10^{-7}	-0.16×10^{-5}

(epoch 1990) were calculated by averaging along the orbit. The results are for $F = \sqrt{X^2 + Y^2 + Z^2}$, X, Y, and Z and are listed in Table 3.6 for both orbits.

3.3.1.1 The External and Disturbance Fields

In contrast to B_i, the external field, B_e, originates from extraterrestrial current sources – primarily the solar wind and the ring current – and exhibits short temporal variations on the order of minutes to days. It can be further classified into two components: the quiet field, Q, and the disturbed field, D.

At the Earth's surface, it is typical practice to divide the Q field, the background field resulting from external currents, into three sources: S_q, the solar-quiet variation; L, the Lunar variation; and M, the magnetospheric variation due to currents in the Earth's plasmasheet. All three components are observed to vary significantly on a day-to-day basis at a given ground station. Typical maximum strengths for these variations in the main field are: $S_q \approx 100 \ \gamma$, $L \approx 10 \ \gamma$, and $M \approx 1 \ \gamma$ ($1\gamma = 10^{-9}$ T). On the average, the quiet external component, $Q = S_q + L + M$, is at least two to three orders of magnitude smaller than the internal main field, B_i. More details about the S_q, L, and M fields are described in Jursa (1985).

The disturbance field component, D, is the variation remaining after the removal of B_i and Q from the total field, B. Thus,

$$B = B_i + B_e = B_i + (Q + D) = B_i + [(S_q + L + M) + D], \qquad (3.3)$$

where D is caused by sudden changes in the interplanetary environment. These changes, resulting in geomagnetic disturbances or storms, are primarily mediated by the solar-wind plasma and the entrapped solar magnetic field. Since Q and D are both time-varying fields, differentiation between them is often difficult. One major characteristic of D is that it usually does not have a simple regularity or periodicity to its temporal variations.

The D field consists of two distinctive components: D_{st}, the geomagnetic storm component; and D_p, the geomagnetic pulsation component. Both D_{st} and D_p are measures of the state of the magnetosphere and, as such, are important to the study of spacecraft interactions. Here, the geomagnetic storms associated with D_{st} are those prolonged disturbance phenomena in the magnetosphere believed to be caused by variations in the solar wind. They are usually preceded by a SSC (a sudden

10–100 γ increase in the horizontal field at the Earth's surface) and are observed globally (Jacobs, 1970). The magnitude of the D_{st} field ranges from a minimum of tens of γ to a maximum of several hundreds of γ depending on the time and location of the recording station. It is proportional to the so-called ring current. This is a ring of plasma encircling the Earth near GEO, formed by the injection of plasma during the geomagnetic storm. The direction of the current is such that the field at the surface of the Earth is depressed, whereas the field at the front of the magnetosphere is increased. In comparison to storms, the geomagnetic pulsations are smaller-amplitude, shorter-period oscillations of the Earth's magnetic field such as in response to the pressure variations in the solar wind.

3.3.2 Low Earth Orbit

Above about 60 km, UV and corpuscular radiation heat and ionize the neutral atmosphere. On the sunlit hemisphere of the Earth, UV, EUV, and X-ray radiation penetrate the neutral atmosphere, ionizing and exciting the molecules present. As the UV/EUV radiation penetrates, it is increasingly absorbed until, by 60 km, it almost completely disappears from the solar spectrum. At the same time, the neutral density and, hence, ionization are increasing. There is thus a fine balance between increasing density and increasing absorption that leads to the formation of ionization layers (principally, the so-called F-layer between 150 and 1,000 km, the E-layer between 100 and 150 km, and the D-layer between 60 and 100 km) that give rise to the mean structure of the ionized component of the atmosphere. The ionosphere is defined as the region from about 60 km to about 1,000 km. It is a transition region from a relatively un-ionized atmosphere to a fully ionized region called the plasmasphere. The plasmasphere is characterized by electron densities of 10^{10} m^{-3} to 10^{11} m^{-3} at an altitude of 1,000 km and then drops to about 10^9 m^{-3} at its outer boundary. The outer boundary of the plasmasphere is called the plasmapause and is characterized by a rapid drop in electron density to 10^5 m^{-3} to 10^6 m^{-3}. The plasmapause is situated nearest the Earth between local times of 2400 and 1800 hours at about four Earth radii and reaches as much as seven Earth radii during local dusk hours. This is called the dusk bulge (Al'pert, 1983). The position of the plasmapause is a function of K_p, being largest when K_p is small (quiet) and smallest when K_p is large.

The peak in the ionospheric density parallels that of the neutral density bulge – occurring approximately two hours after local noon. The ionospheric composition likewise follows that of the neutral atmosphere, varying roughly from NO^+/O^+-dominated in the D-region, to O^+-dominated in the E-region, to O^+/H^+-dominated in the F-region (however, chemical reactions complicate the picture). The ionospheric vertical structure is illustrated in Figure 3.9; for general descriptions of the

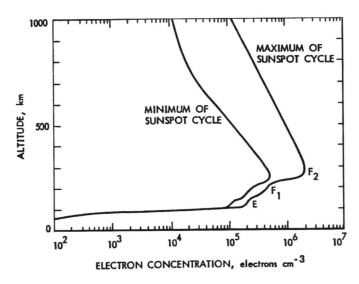

Figure 3.9. Altitude (km) Relative to Electron Density (cm^{-3}) for the Ionosphere.

upper ionosphere, see Whitten and Poppoff (1971), Banks and Kockarts (1973), Al'pert (1983), Kelley (1989), and references therein. Densities reach 10^{12} m^{-3} at the peak in the F-region at about 300 km on the sunlit side. At night, the peak ion density falls below 10^{11} m^{-3} and the composition changes from O$^+$ to H$^+$ above 500 km. Temperatures follow roughly that of the neutral atmosphere, increasing exponentially from a few 100 K at 50–60 km to 2,000–3,000 K above 500 km (i.e., a few tenths of an eV). The electron temperature tends to be a factor of two greater than that of the neutrals, with the ion temperature falling in between. Finally, the density is larger at solar maximum because of the higher UV/EUV fluxes. The ion composition is given in Figure 3.10 for a typical altitude profile at midlatitudes [see also Kasha (1969)].

The principal ionospheric model currently in use is the International Reference Ionosphere (IRI) (Rawer, 1982). This computer model, based primarily on ground-based observations of the total electron content, is the most readily available computer model that gives the electron density (n_e), ion composition (n_i) and temperature (T_e and T_i) as functions of longitude, latitude, altitude (65 to 1,000 km), solar activity (by means of the sunspot number, R), and time (year and local). By compiling and selectively synthesizing all of the reliable experimental data, it uses average profiles as the basis for predictions.

Although the model is limited (it is confined to R values of 150 or less, whereas R values of 200 may occur during solar maximum), it is comprehensive and the most readily available analytical model of the ionosphere (complex computer simulations capable of similar estimates exist but are not readily available). Selected IRI results

Table 3.7. *Input conditions for IRI calculations*

	Max	Min	Mean
Sunspot R	150	10	80
Month	April	July	Sept

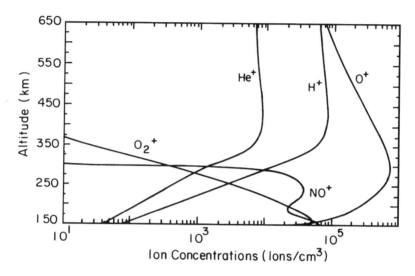

Figure 3.10. Altitude (km) Relative to Ion Composition (cm^{-3}) for the Ionosphere.

are calculated along the *space station* and *EOS* orbits for the values in Table 3.7 for three different input conditions. The parameters were chosen to yield maximum, minimum, and mean plasma densities from IRI for the given orbits. For each orbit, five quantities are presented: n_e, the plasma density (m^{-3}); n_{O^+}, the oxygen ion density (m^{-3}); T_e and T_i, the plasma electron and ion temperatures (both in K); and M, the average ion mass (amu). The results are summarized in Tables 3.8 and 3.9. (*Note:* These are the maximum, minimum, and average values along the given orbit (see Table 1.3), and should not be confused with the input conditions in Table 3.7; the MAX value obtained under the maximum input condition is the maximum value for the mission.)

The IRI model predicts that, unlike the neutral temperature, the electron temperature increases in going from the equator to the pole. Like the neutral density, as discussed earlier, the peak in the electron density is shifted by about two hours from local noon. At *space station* and/or Shuttle altitudes, the ionosphere, primarily because of the corresponding high level of neutral oxygen, is dominated by O$^+$. Even so, the IRI simulation of the *space station* ionosphere shows a complex local

Table 3.8. *IRI results for the* space station *orbit*

Input Table 3.7	Output	n_e (m^{-3})	n_{O^+} (m^{-3})	T_e (K)	T_i (K)	M (amu)
Max	MAX	2.7×10^{12}	1.9×10^{12}	2,653	1,501	11.7
	MIN	3.0×10^{11}	0.7×10^{11}	989	1,019	4.2
	AVE	9.8×10^{11}	5.1×10^{11}	1,497	1,244	7.9
Min	MAX	3.9×10^{11}	2.8×10^{11}	2,531	1,372	11.8
	MIN	2.3×10^{10}	0.5×10^{10}	776	623	4.2
	AVE	1.1×10^{11}	0.6×10^{11}	1,384	988	7.9
Mean	MAX	1.0×10^{12}	0.7×10^{12}	2,643	1,362	11.7
	MIN	8.3×10^{10}	1.8×10^{10}	810	804	4.2
	AVE	3.4×10^{11}	1.8×10^{11}	1,415	1,045	7.9

Table 3.9. *IRI results for* EOS *orbit*

Input Table 3.7	Output	n_e (m^{-3})	n_{O^+} (m^{-3})	T_e (K)	T_i (K)	M (amu)
Max	MAX	1.2×10^{12}	0.9×10^{12}	3,074	2,332	14.0
	MIN	5.8×10^{10}	0.5×10^{10}	905	1,020	2.4
	AVE	3.2×10^{11}	1.8×10^{11}	2,028	1,613	8.4
Min	MAX	1.9×10^{11}	1.4×10^{11}	3,099	2,334	18.0
	MIN	5.1×10^{9}	0.5×10^{9}	847	681	2.2
	AVE	4.2×10^{10}	2.2×10^{10}	2,066	1,519	9.2
Mean	MAX	2.7×10^{11}	1.3×10^{11}	3,078	2,327	14.5
	MIN	2.2×10^{10}	0.2×10^{10}	823	806	2.4
	AVE	8.3×10^{10}	4×10^{10}	1,997	1,531	8.4

time variation with the peak in electron density on the day side. These variations can lead to pronounced changes in the *space station* ram/wake structure.

The same conclusions also apply to the *EOS* orbit. However, for a polar orbit, a spacecraft is subject not only to the cold ionospheric plasma but also to high-energy particles. This modification of the environment is explored in the Section 3.3.3.

3.3.3 Polar Orbits

Unlike a LEO spacecraft, the orbit of a polar platform crosses through the auroral oval regions above latitudes of $\approx 60°$ geomagnetic. In addition to rapid variations in the thermal component of the ionosphere, there is a significant high-energy plasma component not present in low-inclination orbit associated with the auroral zone and, on occasion, the polar cap. The auroral region is a band approximately 6° wide formed into a circumpolar ring. It is centered approximately 3° toward midnight from the magnetic pole and is 25° to 50° in diameter. This is the auroral zone. The visible aurora are the optical emissions produced by the collisions between precipitating energetic particles and the air molecules. The energetic component of the auroral electrons (typical energies between 100 eV and 10 KeV) generates considerable ionization that increases the density of the thermal component. The depth in the atmosphere at which the interaction occurs is dependent on the energy of the particle, with 0.1-keV electrons interacting primarily above 200 km, 1- to 10-keV electrons in the 100- to 200-km region, and higher-energy electrons interacting below 100 km. Unlike the fairly constant UV/EUV fluxes, these corpuscular precipitation events vary rapidly and can increase ionospheric density by orders of magnitude. The occurrence and intensity of the auroras are strongly correlated with magnetic activity, usually measured through K_p.

Two types of auroras have been identified. The diffuse or continuous aurora are barely visible, or subvisible emissions. These are caused by isotropic precipitating electrons of a few keV (Hardy, Gussenhoven, and Holeman, 1985; Hardy, Gussenhoven, and Brautigam, 1989). On the other hand, discrete aurora are bright spatially distinct arcs, bands, rays, and curtains which are caused by precipitating electrons with a Maxwellian energy distribution of ≈ 1 keV plus an acceleration along the magnetic field of several thousand electron volts. The discrete aurora are what are commonly identified as the aurora and are characterized by latitudinally narrow features, some believed to be smaller than a kilometer, and of great longitudinal extent (tens of degrees). The flux of electrons can reach hundreds of $\mu A/m^2$ and can greatly distort the ionosphere. The majority of discrete auroras are found in a belt that depends on local time. Typically, they are found at a magnetic latitude range of 60° to 72° at local midnight and 75° to 77° at noon. These particles can affect the charge balance on surfaces and are important for spacecraft charging.

Both types of auroras are strong functions of magnetic activity. Figure 3.11 shows the hemispheric-number flux of electrons and ions from Hardy et al. (1989) measured as a function of magnetic activity. It can be seen that the electron-number flux increases strongly with magnetic activity. Both the fluence and the flux are critical in determining the effects that these energetic particles will have on the ionosphere, atmosphere, and spacecraft surfaces.

Currently, only a few, very complex thermosphere general circulation models (TGCMs) can adequately model these effects. As it is not yet practical to apply the codes to the problem at hand, a most probable environment based on observation is presented instead.

Here, for the purposes of modeling the effects of the auroras on spacecraft, the distribution in magnetic latitude of the diffuse auroras is assumed to be a Gaussian with a full-width half maximum (FWHM) of 3.2° [MIL-STD-1809 (1991); Jursa,

Table 3.10. *Diffuse auroral energy scale factor and power density*

	Min	Typical	Nominal	Max
P_d (keV $\cdot$ m^{-2} $\cdot$ s^{-1})	1.6×10^{12}	6.2×10^{12}	1.9×10^{13}	7.5×10^{13}
E_m (keV)	0.4	1.5	3.0	9.0

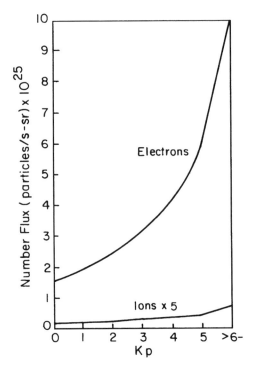

Figure 3.11. Number Flux (Particles $\cdot$ s^{-1} $\cdot$ sr^{-1}) Relative to K_p for the Auroras. From Hardy et al., *Journal of Geophysical Research*, vol. 94, pp. 370–92, 1989. Published by the American Geophysical Union.

1985]. The number flux distribution of the precipitating electrons is assumed to be

$$\phi(E) = \frac{P_d}{2E_m^3} E \exp(-E/E_m), \qquad (3.4)$$

where $\phi(E)$ is the differential-number flux as a function of energy in units of electrons/m^2-s-keV. The energy scale factor E_m and the power density P_d are given in Table 3.10. For this Maxwellian model distribution, the average energy of the precipitating electrons is $2E_m$ and the number flux is $P_d/2E_m$ electrons/m^2-s.

Similarly, Jursa (1985) suggests for the discrete auroras a distribution in magnetic latitude of a Gaussian with an FWHM of 0.1°. The number-flux distribution of the precipitating electrons is given by

$$\phi(E) = \frac{Q_d}{\pi E_s E_g} \exp[-(E - E_g)/E_s], \qquad (3.5)$$

where $\phi(E)$ is the differential-number flux as a function of energy in units of electrons/m²-sec-keV, E_s is a scale factor equal to $0.2E_g$, E_g is the maximum energy of the accelerated electrons, and Q_d is the power flux of the electrons. Typical values for Q_d are 6.25×10^{12} keV/m² · s although it can be as much as 100 times larger than this. The maximum energy E_g takes values in the range 5–18 keV.

3.3.4 The Geosynchronous Plasma Environment

A spacecraft at GEO is on the edge of the plasmapause. Much of the time, a GEO spacecraft will be in a tenuous, relatively cool plasma characteristic of the outer magnetosphere. This region, however, is characterized by sudden injections of high-energy plasma (typically with a mean energy of a few tens of keV) associated with substorms. These events are believed to be the major source of surface charging on GEO spacecraft, making them of particular interest here.

The geosynchronous plasma environment, particularly for the purpose of predicting spacecraft potentials, has been thoroughly reviewed by Garrett (1981) and, more recently, by Garrett and Spitale (1985). Here, only a few of the more salient points from these reviews are mentioned. In particular, the emphasis is on analytical modeling as opposed to the simpler (but more difficult to use) statistical representations and the very complex, first-principle simulations described in the reviews.

In contrast to the situation at LEO where the plasma is sufficiently collisional that the underlying distribution function is Maxwellian, at GEO the plasma is rarefied and collisionless. There is no reason for the distribution function to be close to a Maxwellian. This means that the plasma, in general, cannot be characterized by a single density, drift velocity, and temperature as at LEO. Indeed, to fully characterize the distribution function at GEO, it is necessary either to provide the actual measured distributions or to provide models for the moments of the distribution function. To fully characterize a non-Maxwellian distribution function by moments would require knowledge of all of the moments. Since this is not possible, attempts to model the GEO plasma environment have sought to exhaustively record all of the plasma variations at geosynchronous orbit for as long a time period as possible and at as great a resolution (temporal, energy, pitch angle, mass, and so forth) as possible. Of these, the papers by Garrett and DeForest (1979) [see also Garrett, Schwank, and DeForest (1981a, b)] summarize the ATS-5 and ATS-6 databases whereas the papers by Mullen and his colleagues (Mullen, Gussenhoven, and Hardy, 1986) summarize the *SCATHA* results. In addition, limited summaries are also available from the *GEOS* spacecraft (Geiss et al., 1978; Balsiger et al., 1980; Young, Balsiger, and Geiss, 1982). The ATS and *SCATHA* spacecraft databases have a great advantage over other sources because these spacecraft flew similar plasma instruments and thus provide a "calibrated" source set. These particle detectors measured the plasma variations between approximately 5–10 eV and 50–80 keV. They also provided limited angular data. Roughly 50 complete days at 1- to 10-minute resolution are available for each spacecraft from 1969 through 1980, bracketing one solar cycle.

The statistical models either describe the plasma distribution in terms of the detailed distribution itself or in terms of the moments of the distribution function (DeForest, 1972; Garrett and DeForest, 1979; Garrett et al., 1981a, b). Although the approach of representing the plasma in terms of the detailed distribution function is much to be preferred in scientific applications, it became increasingly obvious over the decade of the seventies that communications satellite operators and designers, the main users of such information, needed simpler models for the purpose of assessing the hazards to their spacecraft.

Subsequent work by DeForest and Garrett detailed the development of a representation of the geosynchronous plasma in terms of the first four moments of the distribution function – number density, number flux (or current density), energy density (or pressure), and energy flux. This technique was found to yield valuable scientific information and, at the same time, to meet the simplicity requirements of the engineers and operators. The leading example of a complete, analytical model capable of such predictive use is that of Garrett and DeForest (1979). This model provided the first four moments of the distribution function in terms of the geomagnetic A_p index and local time for the electrons and ions. Single and double Maxwellian representations of the plasma distribution function were then derived from the moments, which could be used to estimate various spacecraft-charging parameters. The first four moments, M_j, of the electron and ion (assumed to be H^+ for simplicity) distribution functions were defined for isotropic distribution functions as

$$M_1 = 4\pi \int_0^\infty v^0 f_{i,e}(v) v^2 dv$$

$$M_2 = \int_0^\infty v^1 f_{i,e}(v) v^2 dv$$

$$M_3 = \frac{4\pi m_{i,e}}{2} \int_0^\infty v^2 f_{i,e}(v) v^2 dv \qquad (3.6)$$

$$M_4 = \frac{m_{i,e}}{2} \int_0^\infty v^3 f_{i,e}(v) v^2 dv,$$

where M_1 is the number density, M_2 is the number flux, M_3 is the energy density, and M_4 is the energy flux. The double Maxwellian representation assumes that the actual particle distribution function can be written as

$$f_{i,e} = \left(\frac{m_{i,e}}{2\pi}\right)^{3/2} \left[\frac{n_{1_{i,e}}}{(kT_{1_{i,e}})^{3/2}} \exp\left(-\frac{m_{i,e}v^2}{2kT_{1_{i,e}}}\right)\right.$$
$$\left. + \frac{n_{2_{i,e}}}{(kT_{2_{i,e}})^{3/2}} \exp\left(-\frac{m_{i,e}v^2}{2kT_{2_{i,e}}}\right)\right]. \tag{3.7}$$

It is important to emphasize that this representation is for analytical simplicity only. As noted earlier, unlike LEO, where collisions will drive the distribution function to a Maxwellian, at GEO there is no basic reason for the distribution functions to be Maxwellian. However, with this assumption, it can be shown (Garrett and DeForest, 1979) that the number densities N_1 and N_2 and temperatures T_1 and T_2 are uniquely related to the four measured moments M_1, M_2, M_3, and M_4. The relationships are given by

$$M_1 = n_1 + n_2$$
$$M_2 = \frac{n_1}{2\pi}\left(\frac{2kT_1}{\pi m}\right)^{1/2} + \frac{n_2}{2\pi}\left(\frac{2kT_2}{\pi m}\right)^{1/2}$$
$$M_3 = \frac{3}{2}n_1kT_1 + \frac{3}{2}n_2kT_2 \tag{3.8}$$
$$M_4 = \frac{m}{\pi}\left(\frac{2kT_1}{\pi m}\right)^{3/2} + \frac{m}{\pi}\left(\frac{2kT_2}{\pi m}\right)^{3/2}.$$

The power of this approach is shown in Figure 3.12 where measurements from ATS-5 are shown along with a single and two Maxwellian fit. It can be seen that neither the ions nor the electrons can be fit by a single Maxwellian distribution. For the electrons, a single Maxwellian model underestimates the number of low-energy electrons at GEO. For the ions, a single Maxwellian fit underestimates both the low- and high-energy ions. However, a double Maxwellian fit, because it uses four variables rather than two variables, provides a much closer fit to the actual distribution function.

The analytical geosynchronous model in Garrett and DeForest (1979) was based on 10 days of data from the ATS-5 spacecraft between 1969 and 1972. The data were carefully picked to cover a wide range of geomagnetic activity. Care was taken to ensure that a plasma injection event occurred when the spacecraft was at local midnight (in nature, injection events can occur at any time, leading to a complex superposition of many events – thus the desire to isolate the variations associated with a single event). A plasma injection was defined as the sudden appearance of a

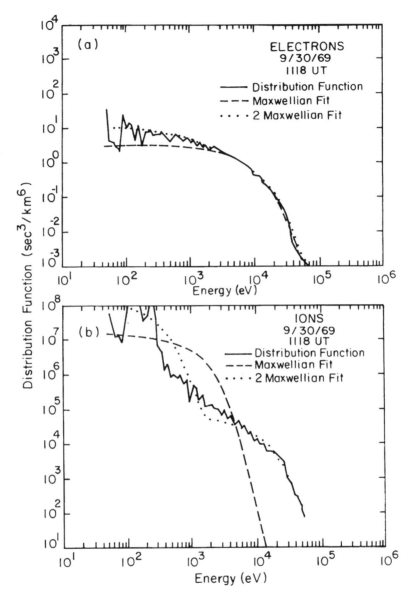

Figure 3.12. Electron and Ion Distribution Functions Observed by ATS-5 Along with Single and Two Maxwellian Fits. Reprinted from Garrett and DeForest (1979) and with the Permission of Elsevier Science Ltd., The Boulevard, Langford Lane, Kidlington OX5 1GB, UK.

dense, relatively high-energy plasma at GEO near local midnight. Local time was measured from local midnight.

The expressions for the four moments of the electron and proton distribution functions for the energy range 50 eV to 50 keV were determined by multiple linear regression fits to a functional form linear in A_p and varying diurnally and

Table 3.11. *Average* SCATHA *environment for electrons*

Parameter	
Number density, M_1 (m^{-3})	$1.09 \pm 0.89 \times 10^6$
Current density, eM_2 (A/m^2)	$0.115 \pm 0.1 \times 10^{-5}$
Energy density, M_3 (eV/m^3)	$3710 \pm 3400 \times 10^6$
Energy flux, M_4 (eV $\cdot$ m$^{-2} \cdot$ s$^{-1} \cdot$ sr^{-1})	$1.99 \pm 2 \times 10^{16}$
n_1 (m^{-3})	$0.78 \pm 0.7 \times 10^6$
kT_1/e (eV)	$0.55 \pm 0.32 \times 10^3$
n_2 (m^{-3})	$0.31 \pm 0.37 \times 10^6$
kT_2/e (eV)	$8.68 \pm 4.0 \times 10^3$

Table 3.12. *Average* SCATHA *environment for ions*

Parameter	
Number density, M_1 (m^{-3})	$0.58 \pm 0.35 \times 10^6$
Current density, eM_2 (A/m^2)	$3.3 \pm 2.1 \times 10^{-8}$
Energy density, M_3 (eV/m^3)	$9440 \pm 6820 \times 10^6$
Energy flux, M_4 (eV $\cdot$ m$^{-2} \cdot$ s$^{-1} \cdot$ sr^{-1})	$2.0 \pm 1.7 \times 10^{15}$
n_1 (m^{-3})	$0.19 \pm 0.16 \times 10^6$
kT_1/e (eV)	$0.8 \pm 1.0 \times 10^3$
n_2 (m^{-3})	$0.39 \pm 0.26 \times 10^6$
kT_2/e (eV)	$15.8 \pm 5.0 \times 10^3$

semidiurnally in local time, LT:

$$M_j(A_p, LT) = (a_{0_j} + a_{1_j} A_p) \{ b_{0_j} + b_{1_j} \cos \left[2\pi \left(LT - t_{1_j} \right)/24 \right]$$
$$+ b_{2_j} \cos \left[4\pi \left(LT - t_{2_j} \right)/24 \right] \}. \tag{3.9}$$

Equivalent values of a_{0_j}, a_{1_j} and so forth are provided in tabular form (Garrett and DeForest, 1979) for the four moments of the electron and ion distribution functions.

The advantages of the model are several. First, it is extremely compact. Second, it gives a reasonable representation of the variations following a substorm injection event at GEO and can be used for rapid estimates of what the geosynchronous plasma will be like if A_p is forecast. Third, the number densities and temperatures derived from the moments [see Garrett and Deforest (1979)] are mathematically consistent with the moments (i.e., the moments or the derived pairs of number density and temperature yield each other uniquely).

This approach applied to the *SCATHA* results gives the average densities and temperatures in Tables 3.11 and 3.12, respectively.

Table 3.13. *Worst-case* SCATHA *environment*

Parameter	Electrons	Ions
Number density, M_1 (m^{-3})	3.0×10^6	3.0×10^6
Current density, eM_2 (A/m^2)	5.0×10^{-6}	1.6×10^{-7}
Energy density, M_3 (eV/m^3)	2.4×10^{10}	3.7×10^{10}
Energy flux, M_4 (eV $\cdot$ m^{-2} $\cdot$ s^{-1} $\cdot$ sr^{-1})	1.5×10^{17}	7.5×10^{15}
n_1 (m^{-3})	1.0×10^6	1.1×10^6
kT_1/e (eV)	600	400
n_2 (m^{-3})	1.4×10^6	1.7×10^6
kT_2/e (eV)	2.51×10^4	2.47×10^4

Although these results are useful for understanding the mean plasma environment to which a GEO spacecraft is exposed, for the purpose of spacecraft design it is often necessary to design against the worst environment that occurs during geomagnetic storms. Geomagnetic storms are widespread disturbances of the geomagnetic field. A storm is normally defined as being in progress when the a_p index is 30 or higher. A geomagnetic storm results when an enhanced stream of solar plasma strikes the magnetosphere, causing changes in the electric currents in the magnetotail. *Sporadic geomagnetic storms* are caused by particle emissions from solar flares and disappearing filaments (also known as eruptive prominences). *Recurrent geomagnetic storms* are caused by discontinuities in the solar wind associated with solar sector boundaries in the interplanetary magnetic field (IMF), or high-speed particle streams from coronal holes. In general, recurrent storms are weaker, show a slower onset, but last longer than sporadic storms. To represent the worst-case environment associated with these geomagnetic storms and their associated substorm injections, a worst-case *SCATHA* environment for spacecraft charging (Purvis et al., 1984) is given in Table 3.13.

A comparison of Tables 3.11 and 3.12 with Table 3.13 indicates that not only does the plasma density increase during a storm but the high-energy part of the distribution function also increases. This results in an increase in current to the spacecraft that will be seen to have a profound impact on the charging of spacecraft surfaces.

3.4 The Radiation Environment

The radiation environment consists of high-energy corpuscular radiation as well as photons. Although this combining of the corpuscular and photon environments may seem somewhat artificial, it will become clear that their interactions with matter are closely related and that their enhancements often occur together. In the following, the natural corpuscular environment is discussed first, the photon environment next, and the man-made environment, which includes examples of both, last.

3.4.1 Energetic Particle Radiation

Here, the high-energy radiation environment is assumed to consist of electrons with energies greater than 100 KeV and protons or heavy ions with energies greater than 1 MeV. The discussion is divided into three sources of radiation:

(1) The Van Allen Belts trapped radiation;
(2) Galactic cosmic rays, which consist of interplanetary protons and ionized heavy nuclei;
(3) Protons and other heavy nuclei associated with solar proton events.

The first two sources are relatively constant or change on long timescales, but the third is highly time dependent.

3.4.1.1 Trapped Radiation

First discovered by J. Van Allen and his collaborators on *Explorer I*, trapped radiation consists principally of energetic protons and electrons, with lesser percentages of heavy ions such as O^+, contained in a toroidal belt(s) around the Earth. This toroid is commonly known as the Van Allen belt(s) (Van Allen, 1971) or the radiation belts and consists of two zones: a low-altitude zone, or inner belt; and a high-altitude zone, or outer belt. The inner belt extends from approximately hundreds of kilometers to ≈6,000 km in altitude and is populated by high-energy (tens of million electron volts) protons and high energy (1–10 MeV) electrons. The outer belt, spanning up to 60,000 km in altitude, is predominately made up of high-energy electrons. A schematic of the radiation-flux contours for the Van Allen belts is presented in Figure 3.13 for electrons and ions. Of particular concern for LEO spacecraft is the extension of the inner radiation belt to low altitudes in the region of the South Atlantic where the Earth's magnetic field is particularly weak (known as the SAA). This can be seen clearly in Figure 3.8 where the field strength drops over the South Atlantic. For GEO spacecraft, even though they are outside the center of the outer belt at $L \approx 7$, they are still subject to fluxes of high-energy electrons. This is well illustrated in Figure 3.14 where the radiation dose on shielded spacecraft is shown relative to distance from the Earth. It can be seen that at GEO the radiation dose is still significant even though it is not deep in the radiation belts. At present, the near-Earth radiation regime is defined in terms of two sets of models – the National Space Science Data Center (NSSDC) AE and AP models. The major characteristics of the radiation environment are summarized for Earth orbits using these AE and AP models with the understanding that they may be subject to revision as new data become available.

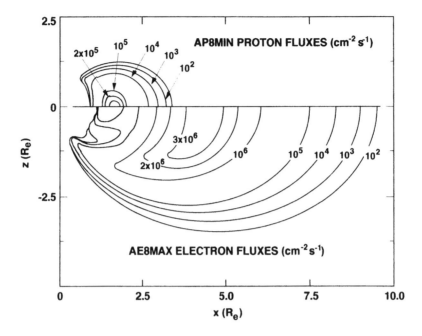

Figure 3.13. Van Allen Belt Radiation Flux for Electrons and Ions.

The detailed mechanism by which particles are entrapped in the belt regions is not well understood nor is the primary source clearly identified. Albedo neutrons are considered to be an important source of the intense proton and electron fluxes observed in the inner belt, and it has been suggested that the the outer belt is due primarily to entrapment of low-energy solar-wind plasma by the geomagnetic field, followed by a subsequent, local hydromagnetic acceleration process. Observations of abundance ratios imply both terrestrial and interplanetary sources.

Once captured, the motions of charged particles in the Earth's magnetic field are governed by the familiar Lorentz force (see Chapter 2). The inner-belt zone, because of the dominance of the Earth's main field, is relatively stable. Most temporal variations in this population occur because, as the solar cycle proceeds and the Earth's neutral atmospheric density at a given altitude changes, the shielding effect of the atmosphere above that altitude varies. Observed counting rates in the inner zone show a factor-of-three variation with time, with electron concentrations being closely correlated with geomagnetic storms. In contrast, the outer belt, which is more influenced by the Earth's highly variable geomagnetic tail, experiences much greater temporal fluctuations. The electron concentration in the outer zone may experience temporal fluctuations as large as a factor of 1,000. Fortunately, most of the physical damage caused by the trapped radiation is largely attributable to the long-term cumulative (or integral) dose received by the spacecraft rather than the instantaneous fluctuations of the radiation.

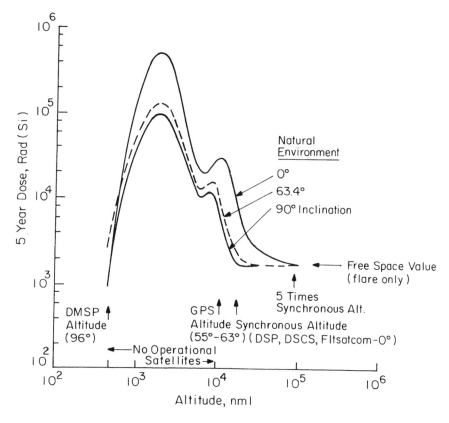

Figure 3.14. Shielded 5-year Radiation Doses in Circular Orbits, with 3 g/cm² Shielding.

The AP8 and AE8 models, developed by J. Vette and his collaborators at Goddard Space Flight Center, are based on compiled data from many different satellites (Jursa, 1985). Here the "P" and "E" in the model names AP8 and AE8 refer to "Proton" and "Electron," and 8 is the version number of the models. For a given set of McIlwain *B-L* coordinates (see Section 2.5.1.1), AP8 and AE8 provide the omnidirectional (i.e., averaged over all pitch angles) fluxes of protons in the energy range from 50 keV to 500 MeV and electrons in the energy range of 50 keV to 7 MeV. Time-dependent variations of the radiation fluxes such as those due to geomagnetic storms or short-term solar modulations are not included in AP8 and AE8. However, the models do differentiate between solar cycle maximum and minimum conditions. For protons, a larger flux occurs at solar-cycle minimum than solar cycle maximum. The situation is reversed for electrons (i.e., higher flux at solar maximum).

The AE and AP models give flux as a function of altitude, latitude, and longitude. The more useful parameter, the fluence, is obtained by time-integrating the flux over the appropriate orbit. Figures 3.15 and 3.16 plot the one-year fluence of electrons and protons for the two extremes in solar activity and altitude for the *space station*

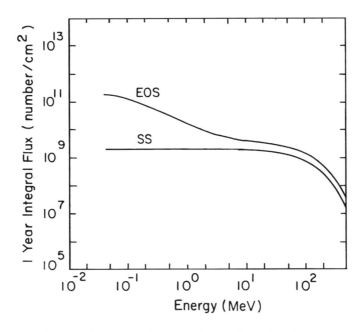

Figure 3.15. Trapped Proton Fluence from AP8-MIN.

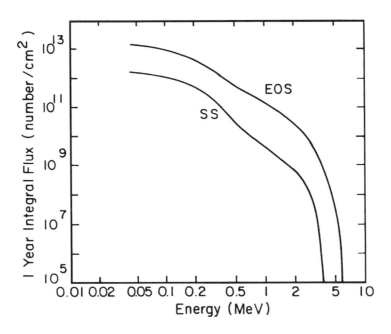

Figure 3.16. Trapped Electron Fluence from AE8-MAX.

and the *EOS* orbits. As both figures show, the *EOS* will encounter higher proton and electron radiation fluences than the *space station* since it passes through higher *L* coordinates and thereby samples more intense radiation regions. The dosage to the spacecraft, the parameter of importance in calculating damage to life or hardware, also will be higher. It and its effects are discussed in Chapter 6.

Recent data from the *Combined Release and Radiation Effects Satellite (CRRES)* has shed new light on the radiation belts. *CRRES* was launched into an 18° inclination geosynchronous transfer orbit on July 25, 1990. The satellite functioned until October 12, 1991. One of the purposes of the satellite was to measure the dynamics of the near-Earth radiation environment and to measure the effects of the space environment on electronics, solar cells, and materials. *CRRES* was active during two large solar events that occurred between March 22, 1991 and May 1, 1991. The event that began on March 22, 1991 was a solar flare larger than 99% of all measured flares. *CRRES* discovered that the influx of solar protons during these events formed a second inner proton belt that still existed when the satellite ceased to function. This second inner belt is being incorporated into updated radiation-belt models.

3.4.1.2 Cosmic Rays

Galactic cosmic rays (GCRs) are primarily interplanetary protons and ionized heavy nuclei with energies from 1 MeV/nucleon to higher than 10^{10} eV/nucleon. Electrons are also a constituent of GCRs, but their measured intensities at energies above 100 MeV are at least one order of magnitude smaller than that of the protons and are usually ignored. Experimental studies indicate that outside the Earth's magnetosphere the cosmic-ray fluxes are isotropic over the entire energy range, suggesting that they are galactic and/or extragalactic in origin. Within the magnetosphere, however, they are not isotropic. For a LEO spacecraft, the Earth's magnetic field deflects many of the lower-energy particles. At these low inclinations, only particles with sufficiently high energy, or rigidity, can penetrate through the magnetic shielding. (See Section 2.5.1.1 for a discussion of of magnetic rigidity.) In the polar regions, particles can enter almost parallel to the magnetic field resulting in a higher and more directional flux, as well as a different energy distribution.

Figure 3.17 shows the measured differential cosmic-ray spectrum outside the magnetic field of the Earth for iron since this is the most damaging of the heavy ions for SEUs. It can be seen that at low energies there is a substantial difference between solar minimum and solar maximum. Figure 3.18 shows the one-year fluence of cosmic rays as a function of energy. Two curves are shown, one for the PEO environment and one for the *space station* orbit. The effects of magnetic shielding are evident. The fluence to the *EOS* extends over a much larger energy range than the *space station* fluence because of the magnetic shielding on the low-inclination *space station* orbit. Table 3.14 displays the relative observed cosmic-ray abundance

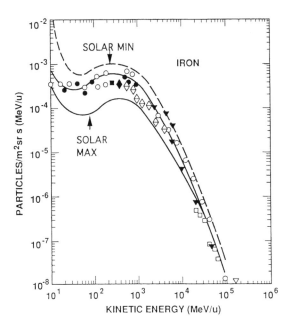

Figure 3.17. Cosmic-Ray Differential Energy Spectrum for Iron (Adams, 1986).

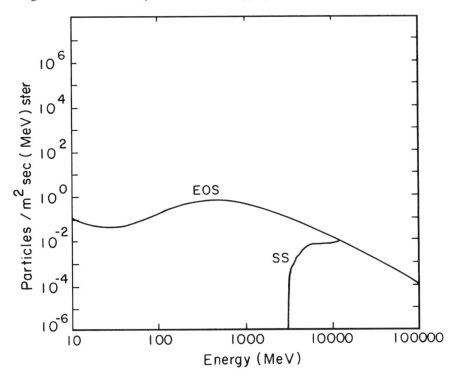

Figure 3.18. GCR Proton Spectrum.

Table 3.14. *Heavy cosmic-ray abundance*

Species	Relative abundance
He	44,700 ± 500
Li	192 ± 4
Be	94 ± 2.5
B	329 ± 5
C	1,130 ± 12
N	278 ± 5
O	1,000
F	24 ± 1.5
Ne	158 ± 3
Na	29 ± 1.5
Mg	203 ± 3
Al	36 ± 1.5
Si	141 ± 3
P	7.5 ± 0.6
S	34 ± 1.5
Cl	9 ± 0.6
Ar	14.2 ± 0.9
K	10.1 ± 0.7
Ca	26 ± 1.3
Sc	6.3 ± 0.6
Ti	14.4 ± 0.9
V	9.5 ± 0.7
Cr	15.1 ± 0.9
Mn	11.6 ± 1
Fe	103 ± 2.5
Ni	5.6 ± 0.6

distribution of the chemical elements in the energy range of 450 MeV/nucleon to 1 GeV/nucleon from hydrogen to the iron group. The LET of these particles, which is a function of both their composition and their energy, is important in determining the SEU rate in electronics and is discussed further in Chapter 6.

3.4.1.3 Solar Proton Events

In close association with solar flares, intense fluxes of high-energy protons are often observed. (*Note:* Solar proton events are often referred to as solar flares; strictly speaking, a solar flare, which involves many processes, may or may not have a classifiable proton event.) Like the GCRs, hydrogen and heavy nuclei in the 1-MeV/nucleon to 10-GeV/nucleon energy range are ejected during a solar proton event. Their intensities are generally a few to several orders of magnitude larger than those of GCRs at these lower energies, depending usually on the size of the solar flare with which they are associated. Models of solar flares and solar

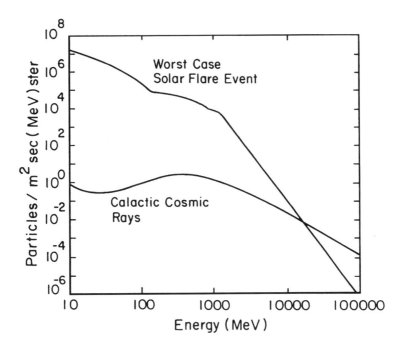

Figure 3.19. Interplanetary Proton Spectrum.

proton events are limited by a lack of sufficient data, particularly in the area of the relative elemental abundances, and the inability to predict solar flare or solar proton occurrence frequency.

The high-energy protons and heavy nuclei associated with proton events, like GCRs, are known to induce SEUs and other malfunctions in digital microelectronic devices. In the past, modelers have used the major proton event that occurred in 1972 in conjunction with mean abundances to obtain a worst-case model. This worst-case solar-flare model has been used widely in worst-case SEU analyses. A comparison between the GCR model and this worst-case solar-flare model is shown in Figure 3.19, where the proton spectra from both models are plotted. As shown, the worst-case solar flare proton flux is five orders of magnitude larger than the GCR flux but becomes "softer" above 10 GeV. As in the case of cosmic rays, entering particles in the polar regions are essentially parallel to the magnetic field, and are not significantly deflected by the magnetic field. At low inclinations, on the other hand, only particles with sufficiently high energy, or rigidity, can penetrate through the magnetic shielding. The spectra of the particles are significantly altered by this process.

Recent work by Feynman et al. (1993) has resulted in the development of an engineering model for the interplanetary fluence of protons with energies between 1 and 60 MeV. The model is based on data collected between 1963 and 1991 and spans several solar cycles. A typical fluence versus probability curve, which is the

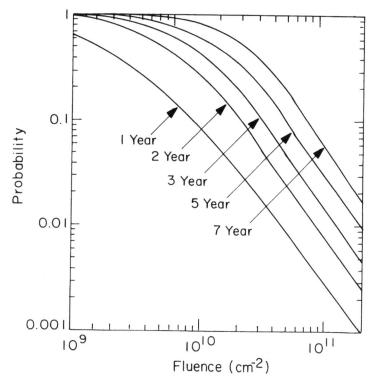

Figure 3.20. Fluence versus Probability Curves for Protons of Energy Greater than 10 MeV for Various Exposure Times. From Feynman et al., *Journal of Geophysical Research*, vol. 98, pp. 13, 281–94, 1993. Copyright American Geophysical Union.

basis of the model, is shown in Figure 3.20. For a given confidence level and mission life, the expected fluence of protons for energies greater than 10 Mev at 1 AU can be determined from this curve. For example, at a confidence level of 95 percent (this means that 5 percent, from Figure 3.20, of missions identical to the one considered will have larger fluences), the fluence of protons greater than 10 MeV for a two-year mission will be $4.9 \times 10^{10}/cm^2$ or less. For protons greater than 4 MeV, at the same confidence level, the fluence for a two-year mission is $1.0 \times 10^{11}/cm^2$, whereas for protons greater than 60 MeV the fluence is $5.0 \times 10^9/cm^2$. The model assumes an inverse-square dependence on heliocentric distance for $r > 1$ AU and r^{-3} for $r < 1$ AU. These fluences are used in subsequent sections in calculations of radiation damage and the probability of SEUs.

3.4.2 Electromagnetic Radiation

The electromagnetic radiation environment is discussed in terms of three frequency ranges:

(1) Electromagnetic radiation at radio frequencies, DC-100 GHz;

(2) Optical, 10^{12}–10^{16} Hz (IR, visible, and UV light);

(3) EUV/X ray, 10^{16}–10^{21} Hz.

3.4.2.1 Electromagnetic Radiation at Radio Frequencies

The natural environment at radio frequencies is highly variable as a function of frequency. Most emissions due to external sources with frequencies below the plasma frequency (ω_{p_e}) at altitudes above the F-region peak of the ionosphere (1–10 MHz) will be severely damped. This implies that interference at frequencies below the plasma frequency will be of local, induced, or magnetospheric origin. These radio noises are generally not significant for spacecraft–environment interactions.

Naturally occurring wave phenomena and spacecraft-induced waves in the near-Earth plasma environment are comprehensively reviewed by Al'pert (1983). Radio noise at frequencies above the 1- to 10-MHz regime are due to galactic EM radiation, solar radiation with both a quiet and an impulsive component extending over a similar frequency range, and terrestrial noise from civilian and military transmitters. The galactic background tends to be broadband in nature, whereas the terrestrial sources are narrowband since they are due to discrete narrowband transmissions.

Solar radio-frequency emissions are quite variable with intense bursts associated with solar flares and related solar events. Solar radio bursts are divided into types (i.e., I, II, III, IV, and V). Type I radio bursts are not believed to be associated with solar flares, whereas Types II and III are associated with plasma oscillations at the plasma frequency. Type II bursts contain the most power of the two. Types IV and V are polarized and associated with solar cosmic rays. Just as with predictions of solar-flare proton fluxes, it is not possible to predict the magnitude or frequency of large radio flares.

3.4.2.2 Visible and Infrared

The visible (3,500–7,000 Å) and IR (0.7–7 μm) portions of the spectrum are dominated by the solar flux. As shown in Figure 3.21, the solar spectral irradiance peaks between 4,500 and 7,000 Å and accounts for most of the solar constant, the total solar energy flux just outside the Earth's atmosphere. It is currently estimated to be $\approx$1,370 W/m^2. Additional sources of visible and IR light that could be of concern include light reflected by the Moon, atmospheric glow, IR radiation from the Earth, and light from auroral displays. Although these sources are far less intense than sunlight, they will, particularly in the Earth's shadow, affect the background for visible or IR instruments. In particular, Earth albedo is a significant contributor to thermal balance in the IR for LEO spacecraft.

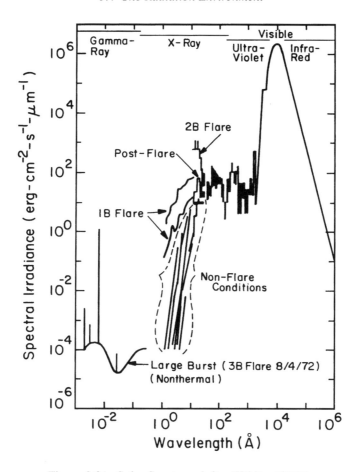

Figure 3.21. Solar Spectrum [after (White, 1977)].

3.4.2.3 UV, EUV, and X Rays

As in the case of the visible and IR portions of the spectrum, the Sun generally dominates the short-wavelength portion of the spectrum in direct sunlight. The shortest wavelengths, 10–100 Å or less, are referred to as X rays. This spectral range contributes to the ionization of the E-region. The spectral region from about 100 to 1,000 Å, called EUV, is related to the photoionization processes of O_2, N_2, and O in the ionosphere and to thermospheric heating. UV radiation is the continuum and line spectrum between roughly 1,000 Å and 3,500 Å. This spectral region contributes to photodissociation, absorption, and scattering processes in the mesosphere, stratosphere, and troposphere. The spectral range from 10 to 1,750 Å is absorbed in the lower thermosphere and affects the production of oxygen atoms and their vertical distribution above the mesopause. The Lyman-alpha line at 1,216 Å plays a major role in the mesosphere through the dissociation of O_2, H_2O, and CO_2

and the ionization of nitric oxide. The spectral region between 1,750 to 2,400 Å leads to the dissociation of O_2 and to ozone production in the mesosphere and stratosphere. Between 2,400 Å and 3,300 Å, the solar irradiance is responsible for the dissociation of ozone and other trace gases that play a role in the stratospheric budget.

Ultraviolet, EUV, and X-ray radiation are not only important to atmospheric and ionospheric dynamics but, through material surface changes and photoelectron emission, are a major environmental factor in space system design at all altitudes. The energy in this spectral range is represented by a solar flux between 10^7 and 10^{10} photons/cm²-s below 1,000 Å. The flux rises almost exponentially to 10^{16} photons/cm²-s between 1,000 Å and 10,000 Å. The flux is not constant but varies in time because of a number of factors, one of which is the solar-cycle variability. This radiation spectrum is also a complex variable of the atmospheric attenuation for a LEO spacecraft that moves in and out of the Earth's shadow.

3.5 The Macroscopic Particle Environment

The macroscopic particle population near a planet comes from two primary sources: interplanetary meteoroids and localized debris. Debris in the Earth's context comprises the waste products of spacecraft operations, the remains of previous satellites, solid particulates from rocket exhausts, paint chips, and so forth, whereas for the other planets, such as Saturn, debris comprises the rings surrounding the planet. These populations are typically very localized (at the Earth, often in the form of shells or streams or, for the other planets, flat, ring shapes). In contrast, the interplanetary environment, characterized by a sporadic background environment and isolated streams, penetrates the entire solar system. Particle sizes of interest vary from micrometers to tens of meters or larger (rocket bodies, the *Skylab*, or perhaps the larger Saturnian ring particles). Typical velocities range from a few kilometers per s to over 50 km/s, making the particles a threat to spacecraft.

Although the primary effect of the particles is mechanical through physical surface damage or optical through scattering of light into sensors, they can indirectly change even the electromagnetic characteristics of space systems. For example, penetration of surface insulation can result in pinholes that expose the underlying conductors to the plasma environment. The subsequent current collection may seriously alter the local surface fields. The ejecta cloud produced by the impact is partially ionized, causing charging and an electromagnetic pulse. The cumulative erosion effects of the impacts will eventually result in failure of exposed insulation, solar array surfaces, and wiring. These and the other effects of the macroscopic particulate environments must be quantified over the projected lifetime of a space system to determine the life expectancy of exposed systems, quantify shielding requirements, and, for the *space station*, establish refurbishment schedules.

The process for modeling the effects of space particulates involves numerous computations. First, a meteoroid or debris model is used to determine a number density versus mass curve, a particle density, and a velocity distribution function (often simply a single, average velocity). Next, these curves are convolved with an appropriate penetration curve (e.g., minimum penetration mass versus velocity) to determine the expected penetration rate. Here, the particulate distributions are addressed. The issues of penetration, cratering, and shielding strategies are outlined in Chapter 7.

3.5.1 The Physics of Macroscopic Particles

The physics of macroscopic particles resembles that of the charged-particle environment as gravity, the controlling force (light pressure and electrostatic forces are ignored here although they are important for smaller or low-density particles), varies as the inverse of the distance between interacting objects. This means that many of the concepts for modeling particle distributions and moments can be applied. In analogy with single-particle dynamics, plasma beams, and Maxwellian distributions for charged particles, the models of the macroscopic environment fall into three groups:

(1) Single-particle dynamics where the trajectories of individual particles are followed. This resembles the plasma physics particle-in-box approach and is used where, in the case of asteroids or large space debris, there are a few (e.g., the $\approx 10,000$ satellites and satellite fragments tracked by NORAD) well-defined "particles."
(2) Organized streams (i.e., meteor streams or newly fragmented Earth-orbiting spacecraft), rings (i.e., Saturn's rings), or shells (small Earth space debris that have been randomized in orbital inclination at a fixed orbital altitude).
(3) The background field environment. This is primarily the so-called sporadic meteors or the zodiacal light. The populations are typically too large to treat as individual particles, and so, averages over the distribution function for various properties are assumed.

To make the similarities to the charged-particle population clearer, consider a particle in space having mass m, position coordinates r (components x, y, z), and velocity $v = dr/dt$ (components v_x, v_y, v_z). The particle can be described as representative of a continuous distribution defined by

$$dN = [H_m dm][g_o(dxdyz)(dv_x dv_y dv_z)], \tag{3.10}$$

where dN is the mean number of particles with mass, position, and velocity in the intervals $(m, m + dm)$, $(x, x + dx)$, ... $(v_z, v_z + dv_z)$. Here, for meteoroids and debris, the dependence on mass m is assumed to reside exclusively in the function

H_m (independent of r and v). It is related to the cumulative mass distribution, H_M, by

$$H_M = \int_\infty^m dm\, H_m, \qquad (3.11)$$

and g_o is a density in position–velocity space like that for a gas or plasma and is independent of m and t. For particulates, g_o can be taken as a function of the constants of motion in a gravity field (e.g., the six Keplerian orbital elements). In particular, it can be shown that g_o can be described for the interplanetary meteoroids in terms of perihelion distance r_1, eccentricity e, and inclination i by assuming that the particles are uniformly distributed in terms of the other elements. Then according to Divine (1993), g_o can be approximated by

$$g_o = (1/2\pi e)(r_1/GM_o)^{3/2} N_1 p_e p_i, \qquad (3.12)$$

where the function p_i depends only on i, p_e depends only on e, and N_1 depends only on r_1 for a specific population. Divine (1993) demonstrates how to derive the required particle concentrations and fluxes from these equations by taking the appropriate moments.

In their simplest implementation, meteoroid (or debris) models assume a functional form for H_M and the moments of g_o (since there must be an integration over phase space). Personnel at NASA Johnson Space Flight Center and their colleagues [Cour-Palais (1969)] and Divine (1993), in particular, have developed detailed interplanetary meteoroid models that determine an equivalent to H_M and approximate the variations in the distribution in terms of radial distance from the Sun, heliocentric longitude and latitude, density, and velocity dependent on the observer's orbit. Both studies identify different characteristic populations. In the case of Divine's model (which incorporates the latest data), there are five distinct populations with densities varying from 0.25 to 2.5 g/cm^3. The older NASA model has only two populations. These are the cometary (0.5 g/cm^3) and asteroidal (3.5 g/cm^3) populations. The latter designations are based on the possible origins of the particles in the model. Here, for simplicity, the interplanetary environment is described in terms of an abbreviated version of the NASA model (Cour-Palais, 1969) because it illustrates the basic components of a meteoroid model and is in widespread use. The reader, however, is referred to Divine (1993) and related papers for the most up-to-date models. Similar remarks hold for the debris environment where the official NASA debris model (Kessler, 1993) will be described. The models to be presented are only meant to be illustrative because the environmental estimates are all undergoing rapid change and reevaluation as data from the *LDEF* in LEO, from hypervelocity impact tests, from various interplanetary sensors, and even the *Clementine* mission continue to alter our understanding. Indeed, the debris environment itself is a very dynamic, rapidly changing distribution in time, and a "snapshot" is all that is realistically possible.

3.5.2 Meteoroid Models

The background interplanetary meteoroids are defined as solid particles orbiting in space that are either of cometary or asteroidal (Cour-Palais, 1969) origin. The spatial volume of interest ranges from 0.1 to 30.0 astronomical units (AU). The mass range is from 10^{-12} to 10^2 g. Knowledge of these particles is based primarily on Earth-based observations of meteors, comets, asteroids, the zodiacal light, and in-situ rocket and spacecraft measurements. The flux versus mass of the particles, the basic quantity required to model the meteoroid environment, is not directly measured over most of the mass range but must be inferred (e.g., from the light intensity of meteors, crater distributions). The ground-based measurements consist principally of photographic and radar observations. These yield fluxes for masses from 10^{-3} g or larger and 10^{-6} to 10^{-2} g, respectively. Observations of the zodiacal light and direct in-situ measurements cover a much smaller mass, typically 10^{-13} to 10^{-6} g. At the other extreme, telescopic observations of asteroids and planetary and lunar crater counts are used to determine the distribution from 50 km and up. As should be obvious, there are data gaps in the assumed distribution. Of most concern to spacecraft is the range from about 10^{-3} to 10 g, because these particles pose a major failure threat, are difficult to detect, and are of sufficient quantity to be a problem. Of concern for pinholes is the mass range from 10^{-3} to 10^{-9} g because these particles have sufficient flux to erode surfaces and sufficient energy to penetrate protective coatings. Meteorites and meteors fall in this range but because of the infrequency of observed impacts and the difficulty of relating the final mass to the original mass, little data are available. Based on these data, the NASA model (Cour-Palais, 1969) divides the observable populations into a low-density (0.5 g/cm^3) cometary and a high-density asteroidal (3.5 g/cm^3) component. Each population has a different characteristic mass, velocity, and angular distribution associated with it. As a result, their importance relative to each other changes relative to the orbit, shielding, or effect. The primary characteristics of the two populations is defined in Sections 3.5.2.1 and 3.5.2.2.

In the discussions that follow, the key function is the flux of meteoroids striking a tumbling, randomly oriented flat surface per unit time. That is,

$$F = \rho_c V'/4, \tag{3.13}$$

where ρ_c = total number of meteoroids above a critical mass, F = particles $\cdot$ m^{-2} $\cdot$ s^{-1} at the spacecraft, and V' is a weighted, or average, relative impact velocity (actually $(\overline{V^{-1}})^{-1}$ in the NASA models). The factor of four comes from the requirement for the particles to impact a flat surface. The problem thus reduces to determining the appropriate number of particles and the average velocity.

3.5.2.1 Cometary Meteoroids

First consider the number density (or mass distribution function) of the interplanetary cometary component as a function of mass. Meteoroids in the mass range of interest (<10 g) are believed to be the solid remains of large water–ice comets that have long since evaporated or broken up because of collisions. The remaining silicate or chondritic material is of very low density (0.16 to 4 g/cm^3). The primary flux of meteoroids inside 1.5 AU is assumed in the model to be made up of these cometary meteoroids because the denser asteroidal meteoroids are assumed to be concentrated in the asteroid belts. On the basis of measurements, the cometary meteor mass distribution at 1 AU in the absence of the Earth is estimated to be (Cour-Palais, 1969)

$$\log_{10} S_c = -18.173 - 1.213 \log_{10} m \tag{3.14}$$

for $10^{-6}g < m < 100$ g, and

$$\log_{10} S_c = -18.142 - 1.584 \log_{10} m - 0.063 (\log_{10} m)^2 \tag{3.15}$$

for $10^{-12}g < m < 10^{-6}$ g, where S_c is the number of cometary meteoroids of mass m (in g) or larger per cubic meter (m^{-3}) at 1 AU.

If a large planet or moon is present, other factors also need to be included. To determine the meteoroid environment at a real Earth (or for the other planets), S_c must be corrected for gravitational focusing (a factor G) and for shielding by the Earth (a factor h). The factor G, which corrects for the gravitational enhancement at a particular distance from the attracting body's center, is expressed as

$$G = 1 + 0.76 \frac{R V_p^2 r_p}{V_e^2 r} = 1 + 0.76 \frac{R_E}{r}, \tag{3.16}$$

where r_p is the radius of the planet (R_E = radius of Earth), V_p = escape velocity from the planet (V_e = escape velocity from Earth), R = distance from the Sun in AU, and r = distance from the planet. The expression on the right is for the Earth.

The correction due to the physical presence of the Earth (or a planet) itself, which shields a randomly oriented spacecraft, can be expressed as

$$h = \frac{1}{2} + \frac{1}{2}\left(1 - \frac{R_E}{r}\right)^{1/2}. \tag{3.17}$$

Multiplying by h has the effect of subtracting out the flux within the solid angle subtended by the shielding body, the Earth in this case.

The full NASA interplanetary cometary model includes corrections to the radial and latitudinal variations in the density. Data indicate a $R^{-1.5}$ variation with heliocentric distance. Similarly, there is a heliocentric latitudinal variation in β (the heliocentric latitude) of the form

$$\exp[-2|\sin(\beta)|].$$

These variations, taken to be independent of each other and the mass distribution, are treated as multiplicative factors for the total density. This gives a final equation [from the Eq. (3.14)] for the density in the range of interest of the form

$$\log_{10}(S'_c) = -18.173 - 1.213 \log_{10}(m) - 1.5 \log_{10}(R) - 0.869|\sin(\beta)|, \quad (3.18)$$

where R is the heliocentric distance (AU).

Combining these factors and treating the Earth as a large spacecraft, the flux is then estimated as the flux to a randomly oriented flat plate by

$$F_{ce} = \frac{\rho_e \overline{V}_c}{4}, \quad (3.19)$$

where $\rho_e = hGS'_c$, F_{ce} = particles/m$^2 \cdot$ s at the Earth, and $\overline{V}_c$ is the "average" relative impact velocity at the Earth (see discussion below; $\overline{V}_c \approx 14.3$ km/s for the Earth at 1 AU). The resulting flux of cometary meteoroids ($\rho = 0.5$ g/cm^3) at 500 km altitude is plotted in Figure 3.22. The flux in this figure is presented as a function of particle diameter, D, where the diameter is assumed to be related to the mass by

$$D = 2\left(\frac{3m}{4\pi\rho}\right)^{1/3}. \quad (3.20)$$

Determination of the proper velocity (actually speed), V_c, to use in the above equation is not straightforward because the cometary meteoroids in interplanetary space vary greatly in their orbital characteristics. In particular, although the majority of cometary meteoroids have prograde orbits, their eccentricities, perihelion distances, and so forth, are randomly distributed over a wide range. To reduce the complexity of the model and make it of practical use to designers, it is assumed in the NASA model that the distribution of eccentricities, inclinations, and the ratios of semimajor axes to distance from the Sun are independent of heliocentric distance. The velocity distribution at 1 AU (i.e., at the surface of the Earth after correction for the gravity of the Earth) then can be assumed for all distances from the Sun. Using this distribution and the known angular variation at the Earth, the relative

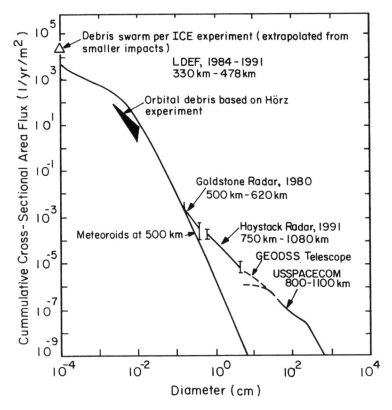

Figure 3.22. Debris and Meteoroid Flux at 500 km. Reprinted from *Advances in Space Research*, vol. 13, D. J. Kessler, Orbital debris environment in low Earth orbit: An update, pp. 139–48, copyright 1993, with kind permission from Elsevier Science Ltd., The Boulevard, Langford Lane, Kidlington OX5 1GB, UK.

collision velocity with a spacecraft in a particular orbit can be computed. An "average" impact velocity is determined by integrating over the velocity distribution (a moment of the velocity distribution function, as discussed earlier). As defined in the NASA model, the weighted velocity for a given power n is

$$\overline{V^n} = \int N(V)V^n dV = \overline{V}^n \delta^{n(n-1)/2}, \qquad (3.21)$$

where V is the relative velocity between the spacecraft and the meteoroid and $N(V)$ is the fraction of meteoroids at V. The δ parameter has been introduced as a method of rapidly approximating different powers of n. It has important consequences for estimating the penetration flux and is defined in detail in Section 7.1.1.5. For now, note that this formulation is equivalent to our standard definition of the moments if $N = \pi g_o V^3$ and $H_M(m) = 4F_{ce}(m)$. For the specific example of the cometary impact flux being considered, the NASA model approximates the "average" relative impact velocity (speed, km/s) as

$$\overline{V}_c(\sigma, \theta, R) = R^{-1/2}(31.29)(1.3 - 1.9235\sigma \cos\theta + \sigma^2), \qquad (3.22)$$

where σ is the ratio of the heliocentric spacecraft speed to that of a circular orbit, and θ is the angle between the spacecraft velocity vector and a circular orbit in the same plane.

3.5.2.2 Asteroidal Meteors

The primary difference, aside from density and velocity, between the cometary and asteroidal components is that the asteroidal component shows a marked heliocentric variation in number density. Visual observations down to masses on the order of 10^{19} to 10^{20} g demonstrate the existence of the well-known asteroid belts between roughly 1.5 and 3.5 AU. It is assumed in the NASA model from the comparative (with respects to the cometary meteoroids) rarity of asteroidal meteoroid falls at the Earth that the lower mass component of the asteroidal meteoroids is similarly confined to the 1.5–3.5 AU range. From laboratory studies of presumed asteroidal meteorites, the density of these particles is assumed to average about 3.5 g/cm^3 – substantially denser then the cometary meteoroids. The mass versus number density curve for the asteroidal meteoroids differs from that of the cometary meteoroids. Unfortunately, this curve is even less well known than the cometary meteoroids and is uncertain by as much two or three orders of magnitude in the mass range of interest to most interplanetary missions [observations (Humes et al., 1974) on *Pioneer 10* and *11* after the publication of the NASA model imply that this population may not even exist at masses below 10^{-9} g and, by extrapolation, may not exist in the mass range of interest to impact studies]. The mass distribution at 2.5 AU is given by the NASA model as

$$\log_{10}(S_a) = -8.23 \qquad (3.23)$$

for $m < 10^{-9}$ g, and

$$\log_{10}(S_a) = -15.79 - 0.84 \log_{10}(m) \qquad (3.24)$$

for $10^{-9}\text{g} < M < 10^{19}\text{g}$.

As for the cometary population, other variations are included as independent variations. These are variations in radial distance from the Sun, $f(R)$, and a variation, j, in heliocentric longitude, l, approximated by

$$j(l, R) = G(R)\cos(l - l_0), \qquad (3.25)$$

where $G(R) = $ heliocentric variation of j and $l_0 = 0°$ (approximately).

An additional latitudinal variation, $h(\beta)$, is ignored here. Again, all variations are assumed to be essentially independent of each other so that the flux is the product of all the components. For the mass range of interest, the resulting equation is

$$\log_{10}(S_a) = -15.79 - 0.84\log_{10}(m) + f(R) + G(R)\cos(l). \tag{3.26}$$

As before,

$$F_a = S_a\overline{V_a}/4, \tag{3.27}$$

where $\overline{V}_a$ (in km/s) is the weighted impact velocity given by

$$\overline{V}_a(\sigma, \theta, R) = R^{-1/2}(30.05)(1.2292 - 2.1334\sigma\cos\theta + \sigma^2) \tag{3.28}$$

for $R = 1.7$ AU;

$$\overline{V}_a(\sigma, \theta, R) = R^{-1/2}(29.84)(1.0391 - 1.9887\sigma\cos\theta + \sigma^2) \tag{3.29}$$

for $R = 2.5$ AU; and

$$\overline{V}_a(\sigma, \theta, R) = R^{-1/2}(29.93)(0.9593 - 1.9230\sigma\cos\theta + \sigma^2) \tag{3.30}$$

for $R = 4.0$ AU.

3.5.3 Space Debris

Spaceflight operations in the Earth's vicinity have led to the creation of an artificial shell of debris around the Earth (Johnson and McKnight, 1991). For spacecraft within 2,000 km altitude of the Earth's surface, this shell of debris poses an impact threat greater than the natural meteoroid environment. The main sources of orbital debris are orbiting spacecraft, fragments from exploded boosters or spacecraft, metal oxides and particulates from solid rocket motors, paint chips, and ejected items from previous missions. These in turn collide with each other creating further debris; for example, the Shuttle *Challenger* had one of its windows pitted by a debris particle, probably a paint chip. There are currently several sets of observational data available for evaluating this growing threat. First, there are ground-based optical and radar observations that form the bulk of the information. These are primarily from the U.S. Space Command orbital element sets for objects of 10-cm diameter and larger, from optical measurements by Massachusetts Institute of Technology for objects 2 cm in diameter and larger, and from debris particle albedo measurements using an IR telescope at the ATMOS/MOTIF, U.S. Space Command radars, and NASA and Space Command telescopes. Second, for particles between 10^{-6} and 10^{-3} cm in diameter, in-situ measurements are available from sample surfaces retrieved from the Solar Maximum Mission (Laurance and Brownlee, 1986) and *LDEF* missions at 500-km altitude. Indeed, recent data have been reported for altitudes out to 120,000 km based on in-situ measurements of micrometer-sized particles on the *Clementine* Interstage Adapter satellite.

Kessler (1991) estimates the cumulative flux of debris on orbiting spacecraft to be given by

$$F(d, h, i, t, S) = kH(d)f(h, S)y(i)[F_1(d)g_1(t) + F_2(d)g_2(t)], \qquad (3.31)$$

where F = flux in impacts per square meter of surface area per year; $k = 1$ for a randomly tumbling surface (must be calculated for a directional surface); d = debris diameter (cm); t = date (year); h = altitude (km), $h < 2000$ km; S = 13-month smoothed 10.7 cm wavelength solar flux, retarded by 1 year from t; i = inclination (degrees); and $y(i)$ = orbital inclination function (Table 3.15).

Here,

$$H(d) = \left\{ 10^{\exp[-(\log_{10} d - 0.78)^2/0.637^2]} \right\}^{1/2}$$

and

$$
\begin{aligned}
f(h, S) &= f_1(h, S)/[f_1(h, S) + 1], \\
f_1(h, S) &= 10^{h/200 - S/140 - 1.5}, \\
F_1(d) &= 1.22 \times 10^{-5} d^{-2.5}, \\
F_2(d) &= 8.1 \times 10^{10} (d + 700)^{-6}, \\
g_1(t) &= (1 + q)^{(t - 1988)} && t < 2011, \\
g_1(t) &= (1 + q)^{23}(1 + q')^{(t - 2011)} && t > 2011, \\
g_2(t) &= 1 + [p(t - 1988)],
\end{aligned}
$$

where p = annual growth rate of intact objects = 0.05; q = fragment growth rate before 2011 = 0.02; and q' = fragment growth rate after 2011 = 0.04.

The average mass density (in g/cm^3) of the debris is approximated by

$$
\begin{aligned}
\rho &= 2.8d^{-0.74} && d > 0.62 \text{ cm} \\
\rho &= 4 && d < 0.62 \text{ cm}.
\end{aligned}
$$

The above equations yield the flux in impacts per square meter of surface area per year. For reference, at low altitudes the relative impact velocity (km/s) can be estimated from the angle α between the impact velocity vector and the spacecraft velocity vector by

$$\overline{V}_d = 15.4 \cos(\pm\alpha). \qquad (3.32)$$

Table 3.15. *Orbital inclination function* $y(i)$

Inclination (degrees)	$y(i)$	Inclination (degrees)	$y(i)$	Inclination (degrees)	$y(i)$
24	0.895	58	1.075	92	1.400
26	0.905	60	1.090	94	1.500
28	0.912	62	1.115	96	1.640
30	0.920	64	1.140	98	1.750
32	0.927	66	1.180	100	1.780
34	0.935	68	1.220	102	1.750
36	0.945	70	1.260	104	1.690
38	0.952	72	1.310	106	1.610
40	0.960	74	1.380	108	1.510
42	0.972	76	1.500	110	1.410
44	0.982	78	1.680	112	1.350
46	0.995	80	1.710	114	1.300
48	1.005	82	1.680	116	1.260
50	1.020	84	1.530	118	1.220
52	1.030	86	1.450	120	1.180
54	1.045	88	1.390	122	1.155
56	1.060	90	1.370	124	1.125

Kessler also gives a detailed velocity distribution function for the velocity. It will not be repeated here but the reader is referred to Johnson and McKnight (1991) or Kessler (1991).

Figure 3.22 presents a comparison from Kessler (1993) of the best orbital debris data and meteoroid fluxes to date. Kessler (1993) defines a critical density as the density of orbital debris at which the population stays constant even as drag removes some particles. This is because lost particles are constantly being replaced by fragments from other particle collisions or by drag from higher orbits. Below 800 km, the current population is below the critical density, whereas between 800 and 1,000 km the current population is above the critical density line. That is, even if launches were ended tomorrow, the latter debris population would continue to grow.

3.5.4 Gabbard Diagrams for Satellite Fragmentation

The background debris environment in terms of particulates orbiting in a uniform shell is defined in the preceeding section. This particulate environment is normally used to determine how many particles will impact a spacecraft in LEO orbit. However, recent data have implied that fragment streams also may play a significant role (at least for micrometer-size particles). Since the particulates in a fragment cloud are subject to rapid change following fragmentation, it is useful to analyze the

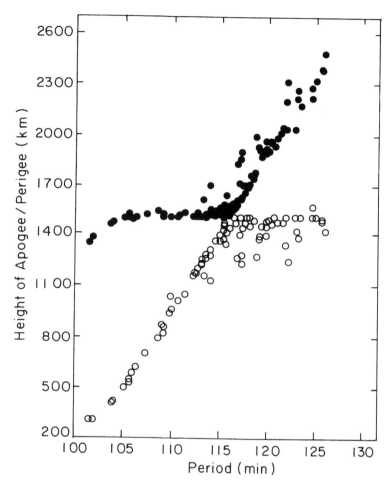

Figure 3.23. Gabbard Diagram for Debris from Satellite 1973-86B (Johnson and McKnight, 1991).

debris patterns and therefore the contribution to the particulate environment when a satellite breaks up. This is done through the use of Gabbard diagrams (Johnson and McKnight, 1991).

The Gabbard diagram plots a satellite's apogee and perigee heights against its orbital period. A circular orbit consists of a point and an elliptical orbit consists of two points that are vertically aligned on the diagram. The debris from the breakup of a satellite that is initially in a circular orbit takes the form of an X on the Gabbard diagram. This is illustrated in Figure 3.23, taken from Johnson and McKnight (1991). The satellite is initially at the center of the X. The debris in the left arm of the X consist of particles that received a velocity impulse that slowed them down relative to the satellite. The debris therefore went onto an elliptical orbit with the apogee the same as the initially circular orbit but with a much smaller perigee. This

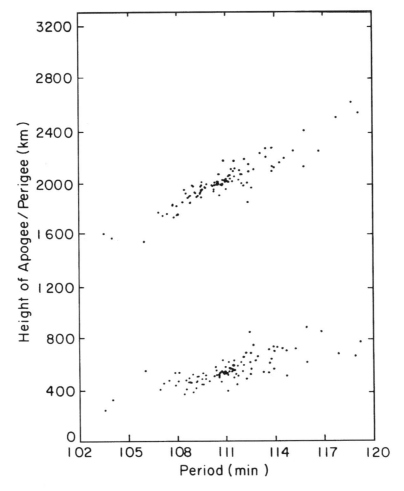

Figure 3.24. Gabbard Diagram for Debris from Satellite 1977-65B. (Johnson and McKnight, 1991).

debris eventually will reenter because it will brush the upper part of the atmosphere. The debris on the right side of the X received an impulse that increased their velocity and therefore went onto orbits that have the same perigee but a much higher apogee. The debris centered in a halo around the center of the X are fragments that received a radial impulse so that the apogee and perigee were affected but the orbital period was relatively unaffected.

In the case of a satellite initially on an elliptical orbit, the debris arms do not cross but spread out at an angle that is a function of the true anomaly at the time of the fragmentation. This is illustrated in Figure 3.24, taken from Johnson and McKnight (1991). The debris spreads out along the two arms and, as before, consists of fragments that receive positive- and negative-velocity impulses.

These diagrams can be used to determine where the debris from a satellite breakup will end up and hence determine the change in the environment on a particular orbit.

3.6 Man-Made Environments

3.6.1 Exoatmospheric Nuclear Detonations

The exoatmospheric (here considered to be above 100 km) detonation of a nuclear weapon produces a variety of radiation environments and effects on spacecraft systems. Here, only a few of the more severe environments are covered. The environments to be considered are those associated with prompt or transient radiation effects and those associated with the long-term effects due to the debris and trapped electrons created by the bomb blast. Electromagnetic pulse (EMP) effects are ignored.

3.6.1.1 Event Morphology

Consider first the overall event morphology of a nuclear detonation. In the general case of a single nuclear detonation above 100 km, the initial effect is the creation of a rapidly expanding plasma. This plasma consists of the nuclear fuel, the ionized remains of the bomb case and delivery vehicle, and the ambient atmosphere. The plasma, because of its charged state is confined to a bubble by the Earth's magnetic field. During the first 100 nanoseconds, the burst emits so-called prompt radiation consisting of gamma rays, X rays, and neutrons. This radiation pulse expands radially, falling off as $1/r^2$ along the line of sight. The delayed effects are associated with the subsequent evolution of the plasma bubble. The hot plasma bubble expands roughly spherically and rises following the initial detonation. Eventually, the plasma cools and contracts, leaving debris trapped on the magnetic-field lines that beta-decays emitting electrons. The debris either is eventually lost to the atmosphere or plates out on spacecraft or other space debris or junk. The high-energy electrons drift eastward in the Earth's magnetic and electric fields creating an artificial radiation belt. This belt is formed in about 6 hours following the initial detonation, after which it begins to decay. Depending on the intensity and location of the belt, it can last from months to years (the effects of belts created by the Starfish and Argus experiments are believed to have been observed through at least one solar cycle of 11 years).

3.6.1.2 The Prompt Environment

The primary effect of the nuclear weapon-produced X rays is energy deposition and ionization. Indeed, approximately 70–80 percent of the energy is in the form of X rays with 1 percent of the energy in neutrons and gamma rays. In the case of low-energy X rays, if the energy deposition takes place sufficiently rapidly, the surface material can be vaporized or actually blown off. Internal stress due to this rapid heating can generate a shock wave in the material that will cause spallation from the interior or exterior surfaces and delamination of dissimilar materials. The intermediate-energy X rays produce effects similar to those of the low-energy electrons, only in the interior of the spacecraft. In addition to these effects, the high-energy X rays can generate photocurrents at the individual device level, causing upset or burnout. X rays are also the major source of system-generated electromagnetic pulse (SGEMP).

Gamma rays, because of their higher energy, cannot be effectively shielded against. They are initially produced in the first 10^{-8} s by the interactions of the neutrons released during the fission and fusion processes with the nuclear reaction and bomb fragments. Gamma rays are subsequently produced by the radiation decay of the nuclear fragments. Although the effects of the gamma rays are similar to those of the X rays and fall off as $1/r^2$, they are, because of their energy, far more penetrating.

Like the X rays and gamma rays, neutrons travel radially outward from the detonation site (i.e., their effects fall off as $1/r^2$). Unlike X rays and gamma rays, they do not travel at the speed of light, but rather at a velocity proportional to the square root of their initial energy. As a result, the highest-energy neutrons reach the spacecraft first, and the lowest-energy neutrons arrive last. The pulse duration at the spacecraft is therefore dependent on the distance between the spacecraft and the detonation; typical values are 10^{-3} s. Neutrons are high energy ($E > 1$ MeV) and have low cross sections because they are not charged, making them very hard to shield against. The much longer duration of the neutron ionization pulse than that of the X rays or gamma rays may be more damaging even though the others may be more intense. Despite this, displacement damage (described later) is the major effect of the neutrons. The neutron spectrum includes a peak at 14.7 MeV that is characteristic of fusion and is not found in fission spectra.

3.6.1.3 Debris Environment

As discussed earlier, ionized debris are created from the vaporization of the weapon's explosion, the casing, the local atmosphere, and any other material in the vicinity of the explosion. Most of the material is constrained to the Earth's magnetic-field lines, but some of the debris can jet radially across field lines away from the detonation. The majority of the magnetically contained debris is lost to the upper atmosphere at the conjugate points. The small amount of debris that is trapped magnetically will produce electrons by beta decay. Spacecraft passing through the debris cloud will become coated with the debris, which can induce a dose (typically small, though the dose rate may be temporarily high) through beta decay, alpha emission, gamma radiation, and neutrons.

Far more important than the debris environment is the trapped electron environment that it creates. Beta decay of the radioactive debris produces large amounts of trapped electrons. This effect, termed the Argus effect after the Argus high-altitude bomb tests of 1958, can lead to the creation of artificial radiation belts. The extent and intensity of the belt so formed is highly dependent on the altitude, yield, and latitude of the detonation. The maximum case corresponds to the so-called saturated nuclear environment in which the total kinetic energy density of the radiation particles just equals that of the Earth's magnetic field (i.e., the radiation can no longer be contained by the Earth's magnetic field if the energy of the particles per unit volume exceeds the magnetic-field energy per unit volume). The energy spectrum of the trapped, weapon-produced electrons is more energetic (or harder) than the natural electron spectrum. The electrons are typically energetic enough to produce secondary X rays – actually, bremsstrahlung – inside even a well-shielded spacecraft. Rate-dependent effects associated with the trapped electrons can be particularly troublesome because their dose rates are often orders of magnitude higher than the dose rates from naturally occurring electron fluxes. Unlike transient effects, which occur only for an extremely short period immediately after the detonation, the duration of the effects due to the trapped electrons is integrated over many orbits.

3.6.2 Nuclear Power Sources

3.6.2.1 Radioisotope Thermoelectric Generators (RTGs)

Interplanetary missions beyond the orbit of Mars typically cannot efficiently utilize solar arrays because the solar energy flux is too low. Rather it has become common practice for missions such as *Voyager*, *Galileo*, and *Ulysses* to employ RTGs. Here, as an example of the induced radiation environment typical of these systems, the *Galileo* RTGs are characterized. It should be remembered, however, that this environment is very device-dependent and will need to be specified for each particular design.

Galileo has two RTGs. These RTGs, called general-purpose heat source (GPHS) RTGs, are basically large cylinders containing golf-ball-sized pellets of plutonium-238 (^{238}Pu). The ^{238}Pu produces heat by emitting alpha particles, and the heat is converted directly into electricity by electric thermocouples. Each RTG contains approximately 11 kg of plutonium or about 137,000 curies (1 curie represents 3.7×10^{10} radioactive disintegrations per s), which gives an intensity of about 7,000 neutrons/s-g at the fuel-cylinder level. The fuel is in the form of pure plutonium oxide (PuO_2) with 0.7 ppm ^{238}Pu and less than 0.5 percent ^{232}U and ^{228}Th. The primary radiation environments of concern are neutrons and gamma rays. Table 3.16 tabulates the normalized differential flux as a function of energy for both RTGs and radioisotope heater units (RHUs) (Hoffman, 1987).

Table 3.16. *Normalized differential gamma and neutron fluxes for the*
Galileo *RTGs and RHUs*

Gamma rays		Neutrons	
Energy width (Mev)	Normal number flux in group	Energy width (Mev)	Normal number flux in group
7–6	1.09×10^{-5}	10–8.55	9.64×10^{-4}
6–5	3.29×10^{-5}	8.55–6.66	4.59×10^{-3}
5–4	9.72×10^{-5}	6.66–5.18	9.93×10^{-3}
4–3	2.96×10^{-4}	5.12–4.46	9.66×10^{-3}
3–2.616	2.65×10^{-4}	4.46–4.04	1.34×10^{-2}
2.616–2.614	1.56×10^{-1}	4.04–3.14	8.73×10^{-2}
2.614–2	1.54×10^{-2}	3.14–2.45	1.76×10^{-1}
2–1.75	2.79×10^{-3}	2.45–1.91	1.57×10^{-1}
1.75–1.5	3.89×10^{-2}	1.91–1.49	1.46×10^{-1}
1.5–1.25	1.75×10^{-2}	1.49–1.16	9.76×10^{-2}
1.25–1.0	2.21×10^{-2}	1.16–0.9	7.6×10^{-2}
1.0–0.75	1.18×10^{-1}	0.9–0.702	5.87×10^{-2}

Note: Data are based on 1.2 ppm ^{236}Pu and a fuel age of 5 years. The flux
will change with fuel age and should be corrected with time.

Table 3.17. Galileo *RHU neutron and gamma radiation distance variation factors to be
used in conjunction with previous table to give absolute values of the flux, fluence, and
dose as a function of distance from the RHU*

Distance from RHU (cm)	Neutrons		γ-rays	
	Peak flux (per cm^2/s)	Fluence (per cm^2)	Peak flux (per cm^2/s)	Dose (Rad (Si))
0	6.1×10^2	1.6×10^{11}	4.7×10^4	5.3×10^3
2	9.5×10^1	2.4×10^{10}	7.3×10^3	8.3×10^2
4	4×10^1	1.0×10^{10}	2.5×10^3	2.9×10^2
6	2.4×10^1	6×10^9	1.5×10^3	1.7×10^2
8	1.2×10^1	3.0×10^9	1×10^3	1.1×10^2
10	8.1×10^0	2×10^9	6.4×10^2	7.2×10^1
15	3.9×10^0	9.8×10^8	3.0×10^2	3.4×10^1
20	2.3×10^0	5.8×10^8	1.8×10^2	2×10^1
50	3.9×10^{-1}	9.8×10^7	3.0×10^1	3.4×10^0

Note: The gamma-ray flux and dose are based on a 5-year old fuel at launch and
must be corrected as a function of time.

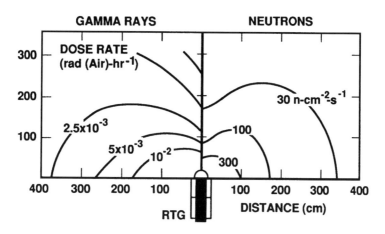

Figure 3.25. Neutron and Gamma Ray Isodose-Isoflux Contours for *Galileo* RTGs.

These normalized values must be multiplied by an absolute value that is a function of both distance and angle relative to the RTG. Figure 3.25 illustrates the neutron and gamma ray isodose-isoflux contours required. The radiation characteristics of the RTG plutonium fuel, because of its short half-life, change rapidly in time. In addition, a factor of two is recommended as a worst-case design margin.

RHUs are intended to provide localized heating within a spacecraft and minimize the use of electrical power. The *Galileo* design is based on a 1-W unit with approximately 34 curies (2.6 g) of plutonium fuel per unit. Each unit was five years old at launch. The PuO_2 fuel pellet is encapsulated within a cladding of platinum–rhodium alloy for fuel containment. Like the RTGs, it is assumed that the plutonium is basically pure ^{238}Pu with less that 1.2 ppm of ^{236}Pu and less than 0.5 percent ^{232}U and ^{228}Th. The normalized fluxes in Table 3.16 are to be multiplied by the values in Table 3.17 to give absolute values. These are given in terms of distance from the RHU in centimeters. As for the RTGs, the gamma-ray values must be corrected by a time factor.

4

Neutral Gas Interactions

For spacecraft in LEO and PEO, the dominant environment is the ambient neutral atmosphere. The neutral gases that make up the atmosphere in this environment form a distinctive structure around the spacecraft and give rise to drag, surface erosion, and spacecraft glow. The neutral gases emitted by the spacecraft itself give rise to contamination on other parts of the spacecraft. In this chapter, these interactions are systematically evaluated. Primary emphasis is on the physics of the flows associated with the interactions.

4.1 Neutral Gas Flow Around a Spacecraft

For a spacecraft in a LEO or PEO, the ambient mean free path for momentum exchange is given by Eq. (2.38). With a typical elastic scattering cross section of $O(10^{-20} \text{ m}^2)$ and with mean densities around the orbit from Table 3.4, the ambient mean free path is of the order of many kilometers. This is illustrated in Figure 4.1 for profiles of the number density, collision frequency, mean free path, and particle speed from the surface to 700 km for the U.S. Standard Atmosphere. Where the Knudsen number [see Eq. (2.39)] satisfies $K_n \gg 1$, the flow of ambient neutral gas around the spacecraft is collisionless. Since from Table 3.4, the gas temperature is typically hundreds to thousands of degrees Kelvin, the thermal velocity is of the order of 700 m/s. For an orbital velocity of 8 km/s, the speed ratio (Section 2.3.2) satisfies $S \gg 1$. Therefore, the ambient neutral gas flow around spacecraft in a near Earth orbit will be collisionless and supersonic. These two observations enable considerable simplification in the equations governing the flow.

Generally, any neutral gas described by a distribution function $f_n(\vec{x}, \vec{v}, t)$ must satisfy the Boltzmann equation (Bird, 1976)

$$\frac{\partial f_n}{\partial t} + \vec{v} \cdot \nabla f_n + \frac{\vec{F}}{m_p} \cdot \nabla_v f_n = C(f_n), \tag{4.1}$$

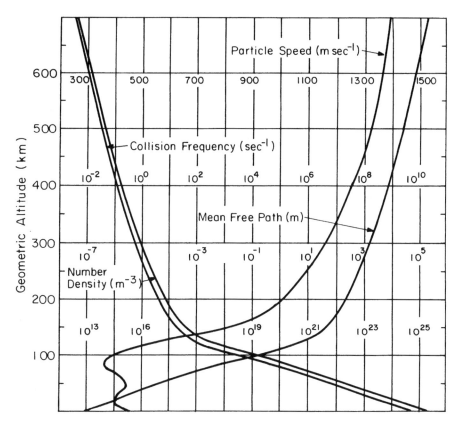

Figure 4.1. Profiles to 700 km of Number Density, Collision Frequency, Mean Free Path, and Particle Speed From the U.S. Standard Atmosphere. From DeWitt et al. (1993). Reprinted by Permission of Kluwer Academic Publishers.

where $\vec{F}$ is the force on a molecule of mass m_p. The complex collision operator $C(f_n)$ includes all of the effects of collisions of the molecules with themselves and with other molecules (Bird, 1976). It is estimated to be $O(\nu f_n)$ where ν is defined in Eq. (2.36) and is the neutral–neutral collision frequency. If the flow of neutral gas is time independent, then a steady state will be reached around the spacecraft where $f_n = f_n(\vec{x}, \vec{v})$. In addition, away from the spacecraft surface there are no significant forces that will act on the gas molecules (on the typical scale L_b, the effect of gravity is so small that it can be neglected). Therefore, the Boltzmann equation for a steady flow of neutral gas around a spacecraft reduces to

$$\vec{v} \cdot \nabla f_n = C(f_n). \tag{4.2}$$

The term $\vec{v} \cdot \nabla f_n$ is $O(v_{\text{th}} f_n / L_b)$ on the spacecraft scale. The ratio of this term to the collisional term gives

$$C(f_n)/\vec{v} \cdot \nabla f_n = \nu f_n / (v_{\text{th}} f_n / L_b) = 1/K_n \ll 1.$$

Therefore, away from the spacecraft, the dominant term in the Boltzmann equation is the convective term, and the Boltzmann equation can be reduced to

$$\vec{v} \cdot \nabla f_n = 0, \tag{4.3}$$

to lowest order in an expansion in the small parameter $1/K_n$.

To introduce the appropriate boundary conditions, $\vec{v}$ can be defined as the velocity of a gas molecule with respect to the spacecraft and $\vec{u}$ is defined as the velocity in a frame fixed to the Earth. Thus, with the orbital velocity $\vec{V}_0$, the two velocities are related by

$$\vec{v} = \vec{u} - \vec{V}_0. \tag{4.4}$$

In the far field, at distances long compared with the body-length scale, the neutral gas in a near Earth orbit is described by a local Maxwellian distribution (see Chapter 3). Hence,

$$f_n(\vec{x} \to \infty, \vec{v}) = n_0 \left(\frac{m_p}{2\pi kT} \right)^{3/2} \exp(-m_p|\vec{u}|^2/2kT), \tag{4.5}$$

where n_0 and T are the density and temperature, respectively, of the neutral gas in the low earth environment far from the body (see Table 3.4 for typical values). In the frame attached to the spacecraft, with the use of Eq. (4.4), the far-field boundary condition is

$$f_n(\vec{x} \to \infty, \vec{v}) = n_0 \left(\frac{m_p}{2\pi kT} \right)^{3/2} \exp(-m_p|\vec{v} + \vec{V}_0|^2/2kT). \tag{4.6}$$

To evaluate the boundary conditions for Eq. (4.3) at the surface of the body, the treatment of Al'pert (1983) is followed. Al'pert solves Eq. (4.3) over an unbounded space but with the presence of the spacecraft taken into account by embedding in the right-hand side of the equation a term of the form $A_n(\vec{r}_s, \vec{v})\delta(F)$, where the surface of the spacecraft is defined by

$$F(\vec{r}_s) = 0, \tag{4.7}$$

and $\vec{r}_s$ is the radial vector to the spacecraft surface and $\delta(F)$ is a Dirac delta function that is nonzero on the spacecraft surface where $F = 0$. Therefore, the equation to be solved over unbounded space with the far-field boundary condition, Eq. (4.6), is

$$\vec{v} \cdot \nabla f_n = A_n(\vec{r}_s, \vec{v})\delta(F). \tag{4.8}$$

The function A_n describes the interaction of the gas molecules with the spacecraft surface such that $\int A_n d^3\vec{v}$ is the variation of the particle flux per unit time caused by the presence of the spacecraft.

To specifically evaluate the function A_n, it is necessary to elucidate what happens when a gas molecule strikes a spacecraft surface (Al'pert, 1983). Three interactions need to be considered:

(1) Specular elastic reflection where the magnitude of the molecule velocity does not change when a gas molecule strikes the spacecraft but the reflection angle to the surface normal equals the incidence angle.
(2) Elastic diffuse reflection or scattering where, again, the magnitude of the molecule velocity does not change on reflection but the molecule can, with equal probability, reflect in any direction away from the surface.
(3) Inelastic reflection or partial accommodation where the magnitude of the molecule velocity is reduced on reflection and there is some distribution of possible angles for reflection.

For a spherical spacecraft of radius R_0 with a radius vector $\vec{r}$ measured from the center of the sphere and with specular elastic reflection (Al'pert et al., 1965), the function $A_n\delta(F)$ is given by

$$A_n\delta(F) = \frac{\vec{r}\cdot\vec{V_0}}{r} f_n(\vec{r},\vec{v})\delta(r-R_0); \quad \vec{r}\cdot\vec{v} < 0$$

$$A_n\delta(F) = \frac{\vec{r}\cdot\vec{V_0}}{r} f_n\left[\vec{r},\vec{v} - \frac{2\vec{r}(\vec{r}\cdot\vec{v})}{r^2}\right]\delta(r-R_0); \quad \vec{r}\cdot\vec{v} > 0,$$

(4.9)

where $\vec{r}\cdot\vec{v} < 0$ means particles moving toward the body and $\vec{r}\cdot\vec{v} > 0$ means particles moving away from the body because they have been reflected. Likewise, for total absorption of the particles at the spacecraft surface, the function $A_n\delta(F)$ is given by

$$A_n\delta(F) = \frac{\vec{r}\cdot\vec{V_0}}{r} f_n(\vec{r},\vec{v})\delta(r-R_0); \quad \vec{r}\cdot\vec{v} < 0$$

$$A_n\delta(F) = 0; \quad \vec{r}\cdot\vec{v} > 0.$$

(4.10)

Before a formal solution to Eq. (4.8) is given, it is valuable to derive on physical grounds what the structure of the solution must be. In Figure 4.2, the density and particle behavior around a model spherical spacecraft are shown schematically. In a collisionless flow, particles only interact with the surface of the spacecraft itself. If the particles are reflected, then the density of particles will be increased in front of the spacecraft. This is known as the ram side compression region. For example, if the particles were all reflected but were thermally accommodated on the surface so that they reflected with a thermal velocity determined by the surface temperature,

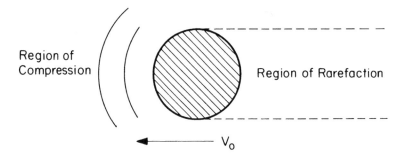

Figure 4.2. Structure of the Gas Flow Around a Model Spherical Satellite in LEO.

from flux balance, the density of particles reflecting from the surface would be

$$n_r = n_0(V_0/v_{\text{th}}),$$

where n_r is the density contribution from the reflected particles and v_{th} is the thermal velocity based on the surface temperature. Hence, the total density right over the surface in the ram side compression region would be

$$n_0 + n_r = n_0[1 + (V_0/v_{\text{th}})] \gg n_0.$$

Far from the body, the spacecraft will look like a point source so that flux balance in spherical polar coordinates gives

$$n_r = n_0(R_0/r)^2 \qquad r \gg R_0.$$

Behind the spacecraft, a wake region will be created because the spacecraft will sweep away the neutral gas. In the absence of collisions, only the random thermal motion of the molecules will allow them to get behind the spacecraft and into the wake region. A molecule will travel a distance $V_0\Delta t$ downstream behind the body in a time Δt. The time that it will take to move laterally a distance R_0 is $\Delta t = R_0/v_{\text{th}}$. Therefore, the length of the wake region behind the spacecraft will be

$$L_w = R_0(V_0/v_{\text{th}}) \gg R_0.$$

For distances behind the body that satisfy $r \ll L_w$, only particles with velocities of $v \approx V_0(R_0/r) \gg v_{\text{th}}$ can get to these locations. Thus, for a Maxwellian distribution of particles, the density just behind the body is given by

$$n \approx n_0 \exp(-v^2/v_{\text{th}}^2) \approx n_0 \exp[-(V_0/v_{\text{th}})^2(R_0/r)^2].$$

The formal solution to Eq. (4.8) (Al'pert et al., 1965) assumes that, since the speed ratio $S \gg 1$, the specific shape of the body has little influence on the wake region for distances long compared with the body linear dimensions. Therefore, if a Cartesian coordinate system is defined such that $z = 0$ corresponds to the center of the body, then the body may be replaced by a disk at $z = 0$ whose cross-sectional area, S_0, corresponds to the maximum cross section of the body (Figure 4.2). In the plane $z = 0$, if the set of points (x_0, y_0) is defined in this plane, then, neglecting the reflected particles, the distribution function in this plane must be

$$f_n(x_0, y_0, 0, v_x, v_y, v_z) = n_0 \left(\frac{m_p}{2\pi kT} \right)^{3/2} \exp\{-[v_x^2 + v_y^2 + (v_z + V_0)^2]/v_{th}^2\}$$

$$v_z < 0 \text{ and } (x_0, y_0) \notin S_0 \quad (4.11)$$

$$= 0; \quad v_z > 0 \text{ or } (x_0, y_0) \in S_0.$$

This clearly satisfies the far-field boundary condition. Behind the spacecraft, Eq. (4.8) is

$$v_x \frac{\partial f_n}{\partial x} + v_y \frac{\partial f_n}{\partial y} + v_z \frac{\partial f_n}{\partial z} = 0. \quad (4.12)$$

This is a hyperbolic equation that has characteristics given by

$$\frac{dx}{v_x} = \frac{dy}{v_y} = \frac{dz}{v_z}. \quad (4.13)$$

The solution for the characteristics is

$$x_0 = x - \frac{v_x}{v_z} z \quad (4.14)$$

$$y_0 = y - \frac{v_y}{v_z} z \quad (4.15)$$

and the solution to f_n behind the body is

$$f_n(x, y, z, v_x, v_y, v_z) = n_0 \left(\frac{m_p}{2\pi kT} \right)^{3/2} \exp\{-[v_x^2 + v_y^2 + (v_z + V_0)^2]/v_{th}^2\}$$

$$v_z < 0 \text{ and } \left(x - \frac{v_x}{v_z} z, y - \frac{v_y}{v_z} z \right) \notin S_0 \quad (4.16)$$

$$= 0; \quad v_z > 0 \text{ or } \left(x - \frac{v_x}{v_z} z, y - \frac{v_y}{v_z} z \right) \in S_0.$$

This distribution function can be integrated over velocity space to obtain the density behind the body. To facilitate the integration, the integration variables can be changed from v_x, v_y, v_z to x_0, y_0, v_z using the relationship in Eq. (4.14). If the density disturbance due to the presence of the body is defined as $\delta n = n(x, y, z) - n_0$, then the density disturbance is

$$
-\delta n = \frac{n_0}{z^2} \left(\frac{m_p}{2\pi kT} \right)^{3/2} \int_{S_0} dx_0 dy_0 \int_{-\infty}^{0} dv_z v_z^2
$$

$$
\times \exp\left\{ -\frac{1}{v_{th}^2} \left[\frac{(x - x_0)^2 + (y - y_0)^2}{z^2} v_z^2 + (v_z + V_0)^2 \right] \right\}. \quad (4.17)
$$

Taking the limit $z^2 \gg (x - x_0)^2 + (y - y_0)^2$ and evaluating the expression over a circular cross section of radius R_0 gives

$$
-\delta n = 2n_0 \exp[-(V_0/v_{th})^2 (r/z)^2] \int_0^{-(R_0/z)(V_0/v_{th})} \rho \, \exp(-\rho^2) I_0 \left(-\frac{r V_0}{z v_{th}} \rho \right) d\rho, \quad (4.18)
$$

where $r = \sqrt{x^2 + y^2}$ and $I_0(p)$ is the Bessel function of zero order with imaginary argument. On the z-axis where $r = 0$, the expression can be simplified to

$$
-\delta n(r = 0, z) = n_0 (V_0/v_{th})^2 (R_0/z)^2. \quad (4.19)
$$

The density contours behind a rapidly moving sphere or ellipsoid are illustrated in Figure 4.3 taken from Al'pert (1983). From Eq. (4.19) it is clear that a density perturbation may exist up to a significant distance behind the body. If the *Space Shuttle* is taken to have a characteristic size $R_0 = 10$ m and velocity ratio $V_0/v_{th} = 8$, then up to 500 m behind the Shuttle, there is a 3 percent or higher perturbation in the ambient density.

The exact structure of the density enhancement in the ram region of the spacecraft depends on the molecular surface interactions as well as the details of the ram side geometry. In general, the calculation can be quite complex. Al'pert et al. (1965) have analyzed the density enhancement for several simple geometries. If there is specular reflection from a spherical surface, then the density enhancement can be calculated from simple geometric considerations combined with the concept of flux conservation. The density in front of a sphere in plane polar coordinates (r, z) is then

$$
n(r, z) - n_0 = n_0 \frac{R_0^2}{r^2} \frac{\sin^2 \theta' \cos^2 \theta'}{1 - (R_0/r) \sin^3 \theta'}, \quad (4.20)
$$

where the angle θ' is determined from

$$
R_0 = 2z \cos \theta' + 2r \sin \theta' - (r/\sin \theta'). \quad (4.21)
$$

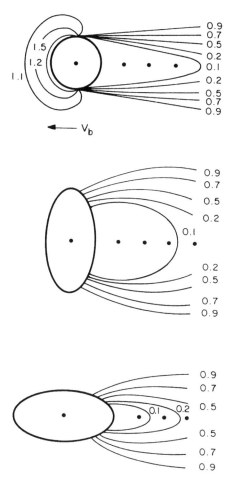

Figure 4.3. Curves of Equal Value of Normalized Density Behind a Rapidly Moving Sphere or Ellipsoid $V_0/v_{th} = 8$. From Al'pert, Copyright 1983. Reprinted with the Permission of Cambridge University Press.

On the axis $r = 0$, the density is given by

$$n(0, z) = n_0 \left[1 + \frac{R_0^2}{(2z - R_0)^2} \right]. \qquad (4.22)$$

At the surface of the sphere $z = R_0$, the total density is exactly twice the ambient density. This can be derived from elementary considerations because each particle moving along the axis is reflected back along the axis. The contours of the normalized density are shown in Figure 4.3 for a sphere. The density enhancement on the ram side extends out to a scale length of the order of the body cross section.

For real spacecraft with convex and concave surfaces that may emit gas, the details of the neutral gas structure have to be calculated numerically. In general, there will be a density enhancement in front of the spacecraft and a long wake with very low densities behind the spacecraft.

The density enhancement in front of the spacecraft requires that the collisionless assumption used to derive the magnitude of the density enhancement be reexamined. In the ambient flow, the mean free path is given by

$$\lambda_{mfp}^{aa} \sim \frac{1}{n_0 \sigma_\infty}.$$

However, the reflected molecules also can collide with incoming ambient molecules. This collision interaction has a mean free path

$$\lambda_{mfp}^{ra} \sim \frac{v_r}{n_0 \sigma_\infty V_0},$$

where the notation λ^{ra} means the collisions between reflected and ambient molecules while λ^{aa} means the collisions between ambient and ambient molecules. If the re-flected velocity v_r is small, as when the molecules give up all their energy to the sur-face and are then reemitted with a thermal velocity characteristic of the surface tem-perature, then the speed ratio $S = V_0/v_{th}$ satisfies $S \gg 1$ so that $\lambda_{mfp}^{ra} = \lambda_{mfp}^{aa}/S \ll \lambda_{mfp}^{aa}$. This suggests that the ambient neutrals may be collisionless with respect to the body, and the neutrals in the ram compression region may be collisional. From this analy-sis, the criterion for the collisionless assumption to be valid everywhere is the more stringent requirement $K_n/S \gg 1$.

4.2 Atmospheric Drag

The impact of the neutral gas molecules in LEO with a spacecraft will transfer energy and momentum to the spacecraft. The spacecraft will feel the momentum exchange as a drag force. Simple momentum conservation considerations give the drag force on a satellite as a force antiparallel to the velocity vector with a magnitude of

$$D = \frac{1}{2} c_d \rho_\infty V_0^2 S, \tag{4.23}$$

where ρ_∞ is the ambient atmospheric neutral density, S is the cross-sectional area of the satellite projected onto the velocity vector, and c_d is the drag coefficient that represents how much the drag deviates from the momentum flux in the ambient free stream.

Atmospheric drag on low Earth orbiting satellites at 150 to 1,000 km is the key parameter in predicting spacecraft lifetime, orbital parameters, fuel requirements, and momentum wheel limits. The satellite cross section is determined by the orientation and configuration of the satellite. This cross section can vary because it is determined by the presence of large solar arrays or antennas. Solar arrays, in particular, can be a problem because they track the Sun and continually change their orientation, resulting in large changes in the effective area of the satellite around an orbit. The atmospheric density depends on a number of factors, such as solar and geomagnetic activity, tides, gravity waves, and winds. Short-term density fluctuations affect orbital position, which complicates tracking and satellite communication. Long-term changes can dominate satellite lifetime, fuel, and reboost requirements. These density variations can be estimated from models such as the MSIS model discussed in Chapter 3.

The drag coefficient c_d depends on the details of how the gas molecules interact with the spacecraft surfaces. To calculate or measure c_d, it is necessary to determine what happens when gas molecules strike a surface. One of the oldest approximations [the Maxwellian approximation (Kogan, 1969)] assumes that, for an incident distribution function f_i, the reflected distribution function of molecules is

$$f_r(\vec{x}, \vec{v}_r, t) = (1 - \tau) f_i[\vec{x}, \vec{v}_r - 2(\vec{v}_r \cdot \vec{n})\vec{n}, t]$$
$$+ \tau n_r \{ m_p/(2\pi k T_r)^{3/2} \exp[-v_r^2/(2k T_r/m_p)] \}, \qquad (4.24)$$

where τ is a free parameter and $\vec{n}$ is the unit normal to the surface. This model states that a fraction $(1 - \tau)$ of the molecules are reflected specularly and a fraction τ are reemitted diffusely from the surface with a Maxwellian distribution based on a temperature T_r. If τ is set to unity, then all impacting molecules are reemitted with an equal probability of leaving at any angle to the surface. The intensity of the molecules being emitted at an angle θ_r is

$$I = \int^{\theta_r} I_0 \sin\theta \, d\theta \, d\phi = 2\pi I_0 \cos\theta_r. \qquad (4.25)$$

This is known as the cosine law and is usually written as

$$dI = I_0 \cos\theta_r \, (d\Omega/\pi), \qquad (4.26)$$

where $d\Omega$ is the differential solid angle.

The possibility that molecules might reflect both specularly and diffusely leads to a definition of the accommodation coefficients for gas molecules interacting with a surface. Assume that a surface at temperature T_s that emits particles with an associated energy E_s is bombarded by a stream of molecules of energy E_i that are reflected at an energy E_r. The energy accommodation coefficient is then defined as

$$\alpha = (E_i - E_r)/(E_i - E_s). \qquad (4.27)$$

If the energies E_i and E_r are associated near equilibrium with temperatures T_i and T_r, then the accommodation coefficient is

$$\alpha = (T_i - T_r)/(T_i - T_s).\tag{4.28}$$

For a monatomic gas, the energy accommodation coefficient can be expressed in terms of the kinetic energy of the molecular streams as

$$\alpha = (v_i^2 - v_r^2)/(v_i^2 - v_s^2),\tag{4.29}$$

where the velocity $v_s^2 = 4kT_s/m_p$.

Two momentum accommodation coefficients can be defined. If a molecule approaches a surface with a normal velocity v_{iN}, tangential velocity v_{it}, and reflects with a normal velocity v_{rN}, tangential velocity v_{rt}, then the normal momentum accommodation coefficient is

$$\sigma = (v_{iN} - v_{rN})/(v_{iN} - v_{sN}),\tag{4.30}$$

where the velocity v_{sN} is the mean normal component of v_s. Since v_s is the equilibrium rebounding velocity distribution at T_s, the corresponding spatial distribution follows a cosine law. Hence,

$$v_{sN} = \int_0^{2\pi} \int_0^{\pi/2} (v_s \cos\theta_r)\cos\theta_r \sin\theta_r\, d\theta_r\, d\phi = 2v_s/3.\tag{4.31}$$

The tangential momentum accommodation coefficient is

$$\tau = (v_{it} - v_{rt})/(v_{it} - v_{st}),\tag{4.32}$$

where v_{st} is the mean equilibrium tangential velocity at T_s. Because the distribution of v_s is a cosine, the average tangential velocity will be zero. Thus,

$$\tau = (v_{it} - v_{rt})/v_{it}.\tag{4.33}$$

The value of the accommodation coefficients is that they are slowly varying functions of the interaction parameters and can be measured easily. They are bounded by zero and unity; $\sigma = \tau = \alpha = 0$ corresponds to specular reflection and $\sigma = \tau = 1$ corresponds to diffuse reflection (this case is also referred to as complete accommodation).

The accommodation coefficients have been measured for many materials used in space (Hurlbut, 1986; Krech, Gauthier, and Caledonia, 1993). For atomic oxygen impacting on nickel and gold, the energy accommodation coefficient has been measured as approximately 0.4 (Krech et al., 1993). In the hyperthermal regime, which is defined as the regime where $v_i \gg v_s$, the accommodation coefficients often are written approximately as

$$\alpha = \frac{v_i^2 - v_r^2}{v_i^2}, \tag{4.34}$$

$$\sigma = \frac{v_{iN} - v_{rN}}{v_{iN}}, \tag{4.35}$$

$$\tau = \frac{v_{it} - v_{rt}}{v_{it}}. \tag{4.36}$$

For LEO where $v_i \simeq V_0 \simeq 8$ km/s and for $T_s = 300$ K, $v_s \simeq 800$ m/s, then $v_i/v_s \simeq 10 \gg 1$. Thus, LEO is well into the hyperthermal regime.

4.2.1 Drag and Lift on a Flat Plate at Angle of Attack β

With the definition of the accommodation coefficients, it is possible to determine the drag coefficient. This is most easily illustrated for a flat plate at an arbitrary angle of attack (Figure 4.4). The momentum imparted per molecular impact in the drag direction, v_i, is

$$f_d = m_p v_i + m_p v_r \cos(\theta_i + \theta_r). \tag{4.37}$$

If this is expanded and the hyperthermal definitions of σ and τ are used, the equation can be rewritten as

$$f_d = m_p v_i [1 + (1 - \sigma) \cos^2 \theta_i - (1 - \tau) \sin^2 \theta_i]. \tag{4.38}$$

The drag force coming from $n v_i A_s \sin \beta$ impacts per s is

$$F_d = n m_p A_s \sin \beta v_i^2 [1 + (1 - \sigma) \cos^2 \theta_i - (1 - \tau) \sin^2 \theta_i] \tag{4.39}$$

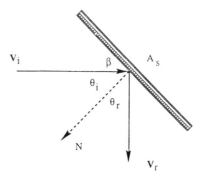

Figure 4.4. Flat Plate at Angle of Attack in a Molecular Stream.

and, using Eq. (4.23), c_d is given as

$$c_d = 2[1 + (1 - \sigma)\cos^2\theta_i - (1 - \tau)\sin^2\theta_i]. \tag{4.40}$$

There are two special cases. The first is specular reflection ($\sigma = \tau = \alpha = 0$); this gives $c_d = 4\cos^2\theta_i$. The second is complete accommodation ($\sigma = \tau = 1$); this gives $c_d = 2$. For the case of diffuse scattering with no specular reflection, the average value of $v_{rN} = 2v_r/3$ and the mean rebounding angle $\theta_r = 0$. If these are substituted in Eq. (4.37), then the force per impact per unit area per second is

$$f_d = m_p v_i \left[1 + \frac{2}{3}(1 - \sigma)\cos^2\theta_i\right]. \tag{4.41}$$

For this type of reflection, the energy accommodation coefficient gives

$$v_r = (1 - \alpha)^{1/2} v_i. \tag{4.42}$$

Hence,

$$c_d = 2\left[1 + \frac{2}{3}(1 - \sigma)\cos^2\theta_i\right], \tag{4.43}$$

or

$$c_d = 2\left[1 + \frac{2}{3}(1 - \alpha)^{1/2}\cos\theta_i\right]. \tag{4.44}$$

The reflected particles also give rise to lift perpendicular to the drag. From Figure 4.4, the momentum in the lift direction per impact is

$$f_L = m_p v_r \sin(\theta_i + \theta_r). \tag{4.45}$$

The lift force is

$$F_L = n m_p v_i^2 A_s \sin\beta \sin\theta_i \cos\theta_i[(1 - \sigma) + (1 - \tau)]. \tag{4.46}$$

With the alternative definition of the lift coefficient,

$$F_L = \frac{1}{2} c_L \rho_\infty v_i^2.$$

This gives

$$c_L = 2\sin\theta_i \cos\theta_i[(1 - \sigma) + (1 - \tau)]. \tag{4.47}$$

Note that for complete accommodation $\sigma = \tau = 1$, the lift coefficient goes to zero.

4.2.2 Drag on a Sphere

Once the flat-plate drag is calculated, the drag on a sphere follows immediately. The differential force on the sphere per area element is the same as that for the flat plate:

$$\frac{dF_D}{dA} = nm_p v_i^2 [1 + (1 - \sigma) \cos^2 \theta_i - (1 - \tau) \sin^2 \theta_i]. \tag{4.48}$$

The differential area element that impacting molecules see is

$$dA = r \sin \theta d\phi (r d\theta) \cos \theta. \tag{4.49}$$

Hence, the total drag is

$$F_D = \int_0^{2\pi} \int_0^{\pi/2} nm_p v_i^2 [1 + (1 - \sigma) \cos^2 \theta_i$$
$$- (1 - \tau) \sin^2 \theta_i] r^2 \sin \theta \cos \theta d\theta d\phi, \tag{4.50}$$

which gives

$$F_D = \left(nm_p v_i^2\right)(\pi r^2) \left[1 + \frac{1}{2}(1 - \sigma) - \frac{1}{2}(1 - \tau) \right], \tag{4.51}$$

and the drag coefficient as

$$c_d = 2 + (\tau - \sigma). \tag{4.52}$$

For a sphere, the range of c_d for $\sigma = 0$, $\tau = 1$ is 3 to 0.5.

In the above analysis, the hyperthermal limit is assumed and the thermal motion of the molecules is ignored. Although this is possible for surfaces that are normal to, or at some angle to, the ram direction, the thermal motion of the molecules cannot be ignored for surfaces that are tangential to the ram direction. In Figure 4.5, a plate tangential to the flow is shown with the rebounding molecules. From the definition of the accommodation coefficients, v_{rt} and v_{it} are related by

$$v_{rt} = v_{it}(1 - \tau). \tag{4.53}$$

The momentum transferred to the surface per molecular impingement and rebound is

$$f_d = m_p(v_{it} - v_{rt}). \tag{4.54}$$

The flux of molecules to the surface that is due to the random motion of the molecules is given in Eq. (2.29). The tangential drag force is then

$$F_{Dt} = m_p(n\bar{c}/4)(v_{it} - v_{rt}). \tag{4.55}$$

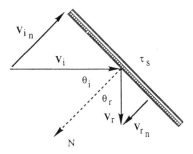

Figure 4.5. Rebounding Molecules from a Plate Tangential to the Ram.

Using Eq. (4.53) and the fact that $v_{it} = V_0$, the tangential drag force is

$$F_{Dt} = m_p(n\bar{c}/4)V_0\tau. \tag{4.56}$$

If the drag force is written as $F_{Dt} = 1/2(c_{dt}\rho_\infty V_0^2 A_t)$, where A_t is the tangential area, then the drag coefficient is

$$c_{dt} = \frac{\tau\bar{c}}{2V_0} = \frac{\tau}{\sqrt{\pi}S}, \tag{4.57}$$

where S is the speed ratio (see Section 2.3.2). The total drag force becomes

$$F_D = \frac{1}{2}c_D\rho_\infty V_0^2 A, \tag{4.58}$$

where A is the surface area projected onto the velocity vector, and the total drag coefficient is

$$c_D = c_d + c_{dt}A_t/A. \tag{4.59}$$

For $\tau = 1$ and $S = 8$, $c_{dt} = 0.07$ and c_d is $O(2)$. Therefore, only if there is a large surface tangential to the flow will the tangential drag add significantly to the total drag.

4.2.3 Effect of Atmospheric Variability on Drag

The general effect of atmospheric drag on the evolution of an orbit of a satellite is the contraction of the apogee altitude. During each orbit, the spacecraft spends increasingly more time at the higher atmospheric densities associated with perigee until the drag becomes so large that the orbit very quickly degenerates and the spacecraft reenters the atmosphere. Note that the drag may not be strictly in the direction tangential to the orbit because atmospheric winds and the rotation of the atmosphere will exert small lateral forces on the trajectory.

One of the most difficult phenomena to calculate for the orbital behavior of a LEO satellite is the exact amount of the drag. This is due to the tremendous variability of the atmosphere at these altitudes. At high latitudes, the density can increase by as much as 400 percent in an hour. This is related to the driving forces of atmospheric tides, solar activity, gravity waves, and so forth. In Chapter 3, the variations in the atmospheric density are discussed in terms of the solar and the geomagnetic indexes. Specifically, the increase of solar ultraviolet, as measured by the $F_{10.7}$ index, increases the amount of energy absorbed by the atmosphere between 120 and 175 km. This energy drives large-scale fluctuations in the atmosphere on a global scale. An increase in the magnetic activity gives rise to streams of high-energy particles that directly deposit energy into the atmosphere at high latitudes and increase the horizontal electric fields, leading to joule heating. These effects all contribute to the sudden large-density fluctuations and high-velocity winds observed over the poles. The effect of the variation in the $F_{10.7}$ index is illustrated in Figure 4.6. The average neutral density at 400 km increases by about a factor of 10 from solar minimum ($F_{10.7} \approx 70$) to solar maximum ($F_{10.7} \approx 230$). As can be seen from the figure, a satellite that would have a lifetime of four years at 400 km and

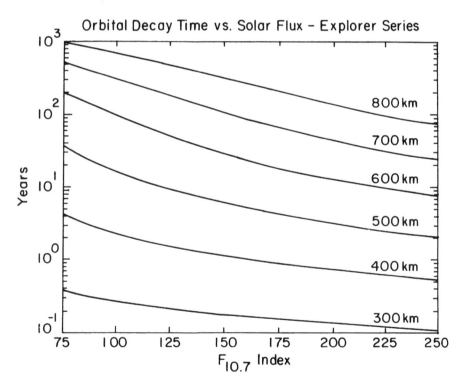

Figure 4.6. Satellite Lifetimes Relative to $F_{10.7}$ for Circular Orbits for Various Initial Altitudes. From DeWitt et al. (1993). Reprinted by Permission of Kluwer Academic Publishers.

solar minimum, would have a lifetime of only six months at solar maximum. There are also seasonal effects on the atmosphere because of its differential heating as the incidence angle of the Sun changes. These long-term effects are well understood and predictable. Tides in the atmosphere arise from solar and lunar gravitational attraction as well as from solar heating, which drives the large diurnal and semi-diurnal global density variations. Gravity waves in the atmosphere, in contrast, are small-scale spatial density and temperature fluctuations of approximately 100 km in dimension. They cause density variations of 30 percent and are driven by a number of sources including auroral particle fluxes, thunderstorms, and mountain ranges. Finally, high-latitude thermospheric winds with velocities up to 1 km/s have been observed. These thermospheric winds can cause drag-induced errors in satellite orbits as well as large cross-track errors for polar orbiting satellites.

4.2.4 Satellite Lifetime and Orbit Determination

For a LEO spacecraft, the orbital lifetime or, conversely, the reboost requirements are determined by the atmospheric density and its long-term fluctuations. As an example, in Table 4.1, the reboost propellant requirements for the *space station* orbit are given as functions of solar activity. Clearly, high solar activity can increase the fuel that must be uploaded to the station by an order of magnitude.

The ability to precisely determine the orbit of a spacecraft is a sensitive function of the atmospheric drag. This need for precise orbit determination often comes from the requirement to accurately point to and communicate with the vehicle. If the in-track and cross-track orbital errors exceed the search area of the tracking and communication equipment, then the satellite can be "lost." Many satellites have limited data storage capacity and require frequent downlinks to provide complete data coverage. When ground terminals are not able to communicate with the satellite, new data and information are written over the older data. This problem has become a critical issue as the number of LEO satellites has increased. Less time is available for searching for and communicating with each individual satellite. For example, during the March 6, 1989 geomagnetic storm, 1,400 satellites

Table 4.1. *Propellant requirements for the reboost of a* space station
in the nominal space station *orbit*

Solar activity	Sunspot number	Propellant required (kg/yr)
Low	50	454
Medium	100	1,362
High	200	4,536

were lost for several days following the huge density variations associated with the storm. The issues discussed in this section thus will become increasingly critical as the number of and pointing requirements for spacecraft increase in the coming decades.

4.3 Contamination

Contamination is the set of processes by which foreign substances adversely affect the performance of space systems and their operations. A contaminant is defined as any foreign substance in front of or on a spacecraft surface. In the context of spacecraft design, a contaminant can be either a molecular effluent or a solid or liquid particle. Although it is sometimes difficult to unambiguously differentiate between gaseous and particulate surface contaminants (e.g., frost and water vapor), gaseous contaminants are assumed here to be individual neutral atoms, molecules, or ions that are either outgassed by materials in space or condense out on spacecraft surfaces. Gaseous contaminants can degrade system performance by condensing on critical optical surfaces or by absorbing light along the line of sight. Typically, the materials used in system construction outgas both adsorbed and absorbed volatile species. The transfer of these outgassed contaminants between spacecraft surfaces is greatly enhanced in the space environment by the low ambient gas density.

Propulsion systems and attitude control systems are significant contributors of contaminants. In particular, although the particulates in plumes are confined to narrow regions in the plume core, some of the gases may flow forward from the nozzle exit plane and impinge on exposed surfaces. As an example of the contaminants produced by propulsion systems, solid rocket motors emit particulate Al_2O_3 and gaseous HCl, H_2O, CO, CO_2, N_2, H_2, and other species; bipropellant motors emit hydrazinium nitrates, H_2O, CO_2, N_2O, N_2O_4, NO_2, NO, CO, CH_4, and H_2; monopropellant thrusters produce N_2H_4, NH_3, N_2, H_2, O, and $C_6H_5NH_2$.

In addition to propulsion systems, spacecraft such as the Shuttle Orbiter periodically release water resulting in high local concentrations of water vapor and ice particles, and gases leak continuously from pressurized cabins. Measurements from the *Shuttle* indicate that the outgassing rate of water is from 0.2 to 0.5 g/s, and the flash evaporator system can release up to 10 g/s of water.

The effects of one common contaminant on IR transmissions are plotted in Figure 4.7. Dioctyl phthalate is one of the outgassed species from polyfluoroethylene used in films and sheets, from polyimide used in hardware and structural materials, and from epoxy used in some kinds of spacecraft paints. As can be seen from the figure, the presence of the material can significantly affect IR sensors over several frequency ranges.

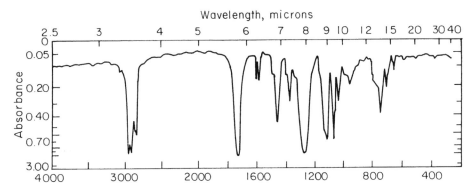

Figure 4.7. IR Absorption Spectrum of Dioctyl Phthalate.

Contamination can occur at all phases of a mission: in the ground segment as the spacecraft is being manufactured, assembled, and mated to the launch vehicle; in the launch and orbital-insertion segments; in the early phases of the orbit; over the long-term mission or during late orbit; and, when applicable, during the postrecovery period following the return of the system to Earth. The contamination processes associated with these phases are equally complicated: particulate contamination; gaseous-contamination spacecraft charging effects; and, recently discovered, spacecraft glow and surface erosion. These processes may significantly degrade performance of systems exposed to the space environment by reducing resolution, light transmission, thermal control, and surface conductivity. These in turn could lead to reductions in mission lifetime and data quality. Small, long-term changes in optical surfaces of a space-based laser, for example, could lead to disastrous results when the system is finally employed. Small changes in the absorptivity of a thermal control surface will reduce its effectiveness and eventually cause loss of thermal control and, ultimately, the satellite. Thus, as spacecraft become more sensitive to environmental effects, the adverse effects of contamination will become steadily more serious.

The contamination phase most open to study and prevention is the effect that ground processes have on the components of the spacecraft system. That is, from the start of manufacture to final assembly, the systems and subsystems are exposed to the deleterious effects of the Earth's surface environment. Effects range from physical contact to actual dirt getting into the spacecraft or on to its surfaces. Although clean-room standards exist and normally are followed at the construction facilities, contamination also can occur during handling, testing, storage, and transportation. Thus, contamination control during the ground phase places severe constraints on space systems from their very beginning! Despite this complexity, comprehensive contamination control programs utilizing air cleanliness standards in the manufacturing facilities, personnel constraints (garment requirements and limited access), and protective measures (dust covers and gaseous purges) coupled with monitoring methods have proven successful in limiting the more serious effects. As reviewed by Jemiola (1980), research is directed at identifying the major contaminant sources in clean rooms and developing methods of relating airborne particulate counts to what actually settles on surfaces. Prebaking of components to ensure that they will meet outgassing requirements, thermal vacuum testing during which released contaminants are monitored, and cleaning of sensitive areas prior to launch are other useful precautions.

The launch and insertion segments of the mission add their own peculiar signatures to the contamination process because payloads will be exposed to contamination induced by aerodynamic heating of the satellite fairing and from thruster firings and staging. For example, significant amounts of mass deposition were observed by the quartz crystal microbalances (QCMs) on the Lincoln Experimental Satellites, LES-8 and *LES-9* during the firing of the Titan second-stage retro-rockets (Figure 4.8). For expendable launch vehicles, the payload tends to be isolated from the launch vehicle, but this is not the case for the *Space Shuttle*, which presents its own peculiar problems. For the Shuttle, not only will individual payloads be exposed to the outgassing from other payloads on the same flight, but they also

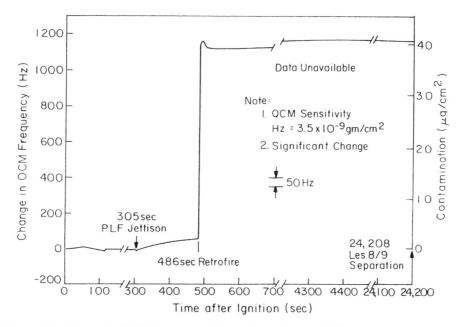

Figure 4.8. Change in QCM Frequency (Contaminant Deposit Thickness) Relative to Time for the Launch of the LES-8 and *LES-9* on a Titan III-C Rocket. From Jemiola (1980). Copyright © 1980 AIAA. Reprinted with Permission.

will be subject to contaminants deposited inside the bay during previous missions. Aggravating the situation is the fact that the vent paths of the Shuttle's major structures pass through the payload areas as do a number of other Shuttle subsystems such as the hydraulic lines – all potential sources of contamination.

During the early *Shuttle* portion of a mission, the launch and insertion phases merge into the on-orbit environment where outgassing is typically a major concern. All materials on Earth accumulate layers of gas molecules. Once the atmospheric pressure is reduced or the temperature is raised, any molecules that are loosely bound to a surface and any molecules with high volatility are released. These include surface-adsorbed species such as water vapor as well as gases distributed throughout the bulk of the material, such as solvents, catalysts, and incompletely polymerized polymers. Outgassing mechanisms include physical desorption from the surface, as well as bulk diffusion of gas to the surface, and chemical reactions on the surface. One of the largest contributors to outgassing on spacecraft is water desorbed from thermal blankets as well as water desorbed from composite materials such as graphite–epoxy structures. Facing surfaces can outgas directly to each other. The physical separation and surface temperature determine how much of the outgassed flux will deposit on the receiving surface. This is modeled analytically in Section 4.3.1. Ultraviolet radiation usually increases the effect of contamination by increasing the deposition rate of molecules onto a surface. In addition, it can photopolymerize the deposition on a surface so that it changes the character of the deposited material. Usually, the deposits become darker (Martin and Maag, 1992).

As evidenced by recent *Shuttle* missions, it takes over a day before atmospheric levels in the *Shuttle* bay drop by 10,000 to near-ambient conditions. During this period, when the bay doors are closed, the mean free path of outgassed molecules approaches that of the dimensions of the *Shuttle* bay. Contaminants can deposit on one surface, be reemitted and deposit on other surfaces inside the closed bay. In this way, contamination in one portion of the bay can reach sensitive surfaces throughout the payload region even if they are shadowed from the source. Even after the doors are opened, the contaminants can redeposit on line-of-sight surfaces. Thruster firings, air and water leaking from the crew's quarters, and the operation of instruments, such as plasma sources, on the *Shuttle* can further degrade operations in this phase.

In addition to the sources of contamination associated with the early stages of the mission, low-level effects occurring over the duration of the mission can become serious. Radiation (particle and electromagnetic) can alter surfaces and transmission properties. Darkening of the cover glass on solar cells is a typical example. The oxygen erosion of surfaces is also a serious problem at *space station* and *Space Shuttle* altitudes for some materials and must be considered very carefully – because the erosion or oxidation of surface materials can produce secondary contaminants,

Table 4.2. *Performance of second-surface mirror radiators*

Spacecraft	Surface temperature (°C)	Initial α	$\Delta\alpha$
NTS-2	−10 to 25	0.15	0.11/yr
SATCOM-1	−10 to 25	0.27	0.02/yr
SATCOM-2	−10 to 20	0.27	0.02/yr
ANIK-B	−10 to 20	0.25	—
DSP	Cold	0.075	0.0012/mo
INTELSATs	Cold	0.08	0.0016/mo
MARISATs	Cold	0.08	0.0012/mo
TIROS/NOAA	0	0.14	0.58/2 yrs

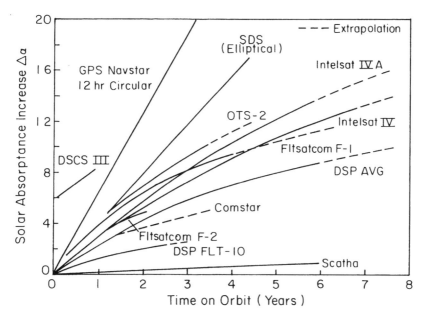

Figure 4.9. On-Orbit Performance of Second-Surface Mirror Radiators. From Paper IAF-92-0336 Presented at the IAF World Space Congress, Washington DC, 1992.

as discussed in Section 4.4. Long-term thermal stresses may crack materials and lead to further outgassing late in the mission.

The thermal control of spacecraft is highly dependent on the thermophysical properties of the surfaces. Many surfaces with desirable properties are susceptible to surface contamination. Surfaces that are particularly sensitive are those with low solar absorbtance or low thermal emittance. These include second-surface thermal control mirrors. In Figure 4.9, the change in radiator absorbance is summarized for some satellites from recent space flights. In Table 4.2, the performance of several

Table 4.3. *Outgassing-species data*

Generic material	Trade name	Outgassed species
Adhesive		
Epoxies		Low-molecular-weight epoxies or polyimides
Circuit boards		
Epoxy glass	Micarta	Styrene
Films and sheets		
Polyterephthalate	Mylar	Glycol dioctyl phthalate
Polyfluoropropylene	Teflon	Carbon Dioxide
Foams		
Polyurethane	Eccofoam	Nitrogen, toluene, styrene
Structural materials		
Polycarbonate	Lexan	Water, mixed hexanes
Sealants		
Silicone	RTV 602	Trimethyl silanol

second-surface mirror radiators is listed. In this table, α is the thermal absorptivity. For the *NTS-2* satellite, the change in solar absorbance led to overheating of the battery radiators, causing battery failure and then spacecraft failure.

With the advent of the Shuttle, post-recovery contamination has become an issue. The heat of reentry and the subsequent bakeout of surfaces, the readmission of atmosphere, ground handling, transportation, and storage are all of potential concern. Covers and storage under nitrogen gas are often used prior to flight to limit these problems, but their reinstitution prior to reentry may be impossible.

Optical systems will be degraded in varying degrees by molecular contaminants. Thermal control surfaces and optical components are particularly sensitive. Molecular contaminants condensing on optical surfaces will, as a function of wavelength and contaminant properties, absorb and scatter the incident light, resulting in reduced signal-to-noise ratio in the optical systems. Generally, the contaminant absorption coefficients are highest in the UV wavelengths, lowest in the visible wavelengths, and moderate to high at IR wavelengths. To limit these effects, screening techniques have been developed to identify materials having the highest outgassing rates. Table 4.3 lists typical materials used on spacecraft and some of their outgassed species. As demonstrated by this table, all types of surface materials can outgas. Currently, good deposition data exist for ground facilities and for on-orbit molecular deposition, volume densities, and species. For the *Space Shuttle*, these data consist of mass spectrometric measurements (Green, Caledonia, and Wilkerson, 1985), witness plates, and QCMs. Effects of engine firings, return fluxes, and outgassing on the *Shuttle* environment have been characterized by Wulf and von Zahn (1986). These measurements are discussed in Section 4.3.3.

4.3.1 Modeling of Contamination

In the preceding section, several examples of the deleterious effects of contamination were discussed that show how serious contamination can be to the design and operation of a spacecraft. In this section, the process of modeling contamination and its effects is described. Modeling of contamination on spacecraft is conventionally divided into four steps:

(1) determination of the source of the contamination,
(2) transport of the contaminants,
(3) accommodation of the contaminants on the surface, and
(4) determination of the effects of the contaminants on the surface and on spacecraft operations.

Consider first the sources of contamination on a spacecraft: thruster exhausts, overboard vents, leaks from pressurized volumes, and outgassing of spacecraft materials.

Thruster plumes have an enormous potential for contamination because of the large amount of mass released during a thruster firing. They are examined in Section 4.3.2, along with the effects from large overboard vents and leaks. For small leaks and vents, the analysis is the same as for material outgassing and is considered below.

All materials outgas when first placed in the ultra-low-pressure environment of space. The rate of outgassing is a strong function of material type, previous processing, material temperature, and the exposure time to the space environment. There are three main mechanisms by which materials outgas:

(1) desorption of gases physically or chemically adsorbed on material surfaces;
(2) evaporation of gases in solution in the material; and
(3) evaporation or sublimation of the material itself.

Desorption of surface molecules and evaporation of dissolved gases account for the majority of outgassed contaminants for most spacecraft materials. Generally, the outgassing rate is a strong function of material type and processing history. The material's exposure to ambient gases is important because it determines what total volume of gas may have accumulated in the material prior to exposure to the space vacuum. For outgassing from materials with a limited source of dissolved and adsorbed gases, the emission rate decreases slowly over time as the quantity of gas in the material decreases. It has been observed empirically that metals tend to outgas according to a time dependence t^{-1}, producing gases such as H_2O, CO_2, and O_2. Glass materials tend to follow a $1/\sqrt{t}$ dependence, producing mainly H_2O and CO_2. Typical initial loss rates for large space components such as radiators are of the order of 10^{20} molecules/s.

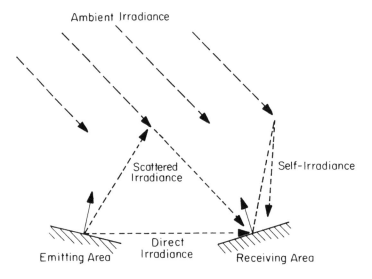

Figure 4.10. Molecular Irradiance Paths.

Once the sources of the outgassing have been identified, the next step is the transport of the contaminants. The density of the contaminants from outgassing is sufficiently small that the molecules move ballistically or have a small number of collisions before striking another surface. This divides the discussion of the contaminant efflux to a surface into a discussion of three paths (see Figure 4.10): (1) direct irradiance, (2) scattered irradiance, and (3) self-irradiance. In direct irradiance, the receiving surface "sees" the emitting surface and molecules move ballistically from the emitting surface to the receiving surface. For scattered irradiance, the molecules are desorbed by the emitting surface and scattered by another surface or by other molecules into the receiving surface. Self-irradiance occurs when molecules are desorbed by a surface and scattered back to the same surface.

Direct irradiance is somewhat simple to analyze and avoid. The density of molecules just above the receiving surface is

$$\rho = \rho_s(\Omega/2\pi), \qquad (4.60)$$

where ρ_s is the density just above the emitting surface and Ω is the solid angle subtended by the receiving surface measured from the emitting surface. The simple solution is not to place any materials with substantial outgassing rates in the line of sight of surfaces that are sensitive to contamination.

The analysis of scattered or self-irradiance is complex and depends on the details of the spacecraft geometry and the ambient environment. Generally, these analyses are done for actual spacecraft by Monte Carlo simulations (Bird, 1976). However, it is possible to show that these types of contamination are mainly an issue for low-orbiting spacecraft. This is demonstrated as follows.

If a spacecraft is modeled as a sphere of radius R and the desorbed molecules move out radially until they collide with the ambient and are stopped, then the change in the flux of desorbed molecules at a distance x from the surface is

$$d\Gamma_d = -\Gamma_d(dx/\lambda_{\text{mfp}}), \tag{4.61}$$

because the probability of a molecule being stopped is proportional to the distance it has to travel in units of the mean free path. If the initial outgassing rate on the surface is N_d and the thermal velocity of the desorbed neutrals is v_d, then the desorbed neutral density is

$$n_d = \frac{N_d \exp(-x/\lambda_{\text{mfp}})}{4\pi(R+x)^2 v_d}. \tag{4.62}$$

The mean free path of a desorbed neutral in terms of the mean free path at infinity is

$$\lambda_{\text{mfp}} \simeq \frac{v_d}{V_0 + v_d}\lambda_\infty. \tag{4.63}$$

Hence, the flux of desorbed neutrals is

$$\Gamma_d = \frac{N_d \exp\{-[(V_0 + v_d)/v_d]x/\lambda_\infty\}}{4\pi(R+x)^2}. \tag{4.64}$$

The flux of molecules that are scattered back is determined from

$$\Gamma_b = -\int_0^\infty d\Gamma_d \approx \frac{N_d}{4\pi\lambda_{\text{mfp}}R}. \tag{4.65}$$

The ratio of the return flux to the desorbed flux at $x = 0$ is

$$\frac{\Gamma_b}{\Gamma_d} = \frac{R}{\lambda_\infty}\frac{V_0 + v_d}{v_d}. \tag{4.66}$$

This ratio is shown in Figure 4.11 for typical ambient parameters and for a desorbed neutral thermal velocity of 400 m/s. Based on the parameters in the figure, the self-irradiance becomes very small above 600 to 800 km and can safely be ignored at those altitudes and above.

Once a flux of contaminant molecules reaches the receiving surface, molecules may either be reflected or may adhere to the surface. If they reflect, perhaps with some energy or momentum accommodation, then they will behave as if they were outgassed from the reflecting surface and may again be back scattered to the spacecraft by the mechanisms discussed. Contaminants that adhere can either condense or be adsorbed onto the surface. Condensation occurs when the pressure exerted by a contaminant gas over a surface exceeds the saturated vapor pressure of the gas at

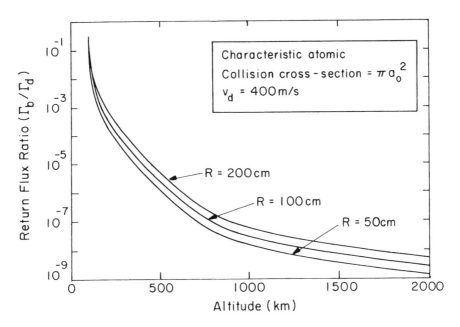

Figure 4.11. Returning Flux Relative to Altitude.

the surface temperature of the material. Condensation can be a very serious problem because it easily forms a thick layer on a surface. It is usually avoided on spacecraft surfaces by using materials that emit a very small fraction of volatile condensible material (VCM). The current standard is that materials used on a spacecraft have less than a 0.1 percent mass loss of VCM when exposed to vacuum for one hour at 125°C.

After VCMs, the main source of deposition on spacecraft surfaces is adsorption of individual molecules. An adsorbate forms because of surface attraction between individual atoms of the substrate and those of the contaminant. The degree of adherence of any individual particle depends on the gas species, the surface temperature, the composition of the substrate, and the amount of surface coverage. As a monolayer is completed, the likelihood decreases that additional contaminant molecules will stick because they will not see any substrate molecules. If a surface has both condensate and adsorbed layers on it and the fraction of the surface that is covered by condensate is f_b, then, quite generally, the net rate at which mass builds up on a surface can be written as

$$\dot{m} = \dot{m}_i[S_a(1 - f_b) + S_b f_b] - [\dot{m}_b f_b + \dot{m}_d(1 - f_b)], \qquad (4.67)$$

where S_b is the fraction of the incoming molecules that stick to the condensate, S_a is the fraction that stick to the adsorbate, $\dot{m}_i$ is the incoming mass flux of contaminants, $\dot{m}_b$ is the bulk reevaporation rate, and $\dot{m}_d$ is the desorption rate. The sticking coefficient S_b is a property of the condensing species and is usually close to unity. The bulk reevaporation rate $\dot{m}_b$ is related to the vapor pressure of the condensing species by the Langmuir equation

$$\dot{m}_b = m_s \frac{p_v(T)}{\sqrt{2\pi m_s kT}} A, \tag{4.68}$$

where m_s is the molecular mass of the condensing species, T is the temperature of the surface, A is the area of the surface, and $p_v(T)$ is the equilibrium vapor pressure of the gas above the surface. The equilibrium vapor pressure is usually measured. However, from Eq. (4.68), it can be seen that gases with a low vapor pressure will accumulate readily on a surface, whereas gases with a high vapor pressure will not. The terms S_a, f_b, and $\dot{m}_d$ are highly dependent on the interaction between the gas and the surface as well as the temperature and must be measured for each situation. The term f_b also depends on the mobility of the condensate on the surface. The desorption rate $\dot{m}_d$ usually has the form

$$\dot{m}_d = A \frac{m_s w_d}{\tau_d} \exp\left(-\frac{E_d}{kT}\right), \tag{4.69}$$

where w_d is the surface density of the adsorbate layer in particles per unit area, τ_d is a characteristic lattice vibration time, and E_d is an activation energy for desorption. For molecules that are physically adsorbed, $E_d \leq 1$ eV; for molecules that are chemically adsorbed, $E_d \approx 3-5$ eV. The physical content of Eq. (4.69) is that particles are desorbed when they acquire enough energy to break the bonds on the surface and that the characteristic rate at which they acquire such an energy is related to the lattice vibration time, which is of the order of 10^{-13} s.

The range of outgassing parameters and effects can be illustrated by considering an actual example. Rault and Woronowicz (1993) analyzed the Upper Atmospheric Research Satellite (*UARS*) for contamination of the Halogen Occultation Experiment (HALOE) infrared telescope. The experiment was designed to measure the concentration of trace gases responsible for oxygen depletion, and so it was extremely susceptible to contamination. The contamination on the instrument was calculated with a three-dimensional direct-simulation Monte Carlo model for the *UARS* geometry. The Monte Carlo simulation allows the calculation of the direct irradiance as well as the scattered and self-irradiance. The simulation was performed for a *UARS* orbit of 600 km with a nominal atmosphere of atomic oxygen at a temperature of 1,000 K and a concentration of 6×10^{12} molecules/m^3. The mean free path in the undisturbed freestream was $\lambda_\infty = 400$ km. The satellite velocity was $V_0 = 7.5$ km/s, and the satellite surface temperature was 300 K. Two types of outgassed species were modeled. One species was taken to have a molecular weight of 100 g/mole and was vented from a slit in the satellite on the side opposite to the

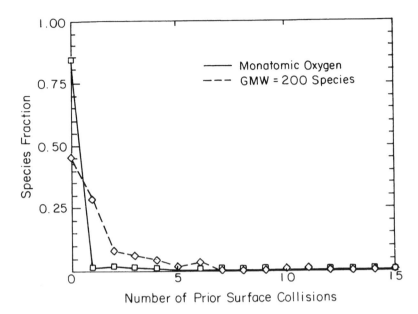

Figure 4.12. Number of Surface Collisions Encountered by Particles Prior to Striking the HALOE Aperture. From Rault and Woronowicz (1993). Copyright © 1993 AIAA. Reprinted with Permission.

HALOE at a rate of 2×10^{18} molecules/m²-s. The other species was taken to have a molecular weight of 200 g/mole and to be emitted from all exposed surface at rates of 9×10^{13} to 4×10^{15} molecules/m²-s. The molecules were allowed to collide with each other as well as with the ambient oxygen [a variable hard-sphere model was assumed (Bird, 1976)]. Figure 4.12 is a plot of the number of collisions that each particle encountered in reaching the HALOE location. Of the molecules reaching the HALOE location, 98 percent were atomic oxygen but only 2 percent were of molecular weight 200. Of the impacting molecules reaching the instrument, 80 percent had suffered no collisions. This suggests that the easiest way to protect the instrument is to point it away from the ram and ensure that no part of the spacecraft is in its field of view.

4.3.2 *Thruster Contamination*

The potential for contamination from thruster firings can be very large. An upper stage such as the inertial upper stage (IUS) has a propellant load of 10,000 kg, all of which is exhausted into space during a thruster firing. To substantially degrade the performance of a space radiator, as little as 0.1 g of material deposited on the surface will suffice. Only one molecule in 10^8 need deposit on a radiator from the plume to significantly reduce the radiator's performance.

Thruster exhaust products may reach spacecraft surfaces by three mechanisms. The first mechanism is differential charging of the plume and spacecraft, which causes electrical attraction between the exhaust products and the spacecraft surfaces. This is usually small for chemical thrusters, despite the spacecraft having a substantial charge, because there is very little ionization in the plume of a chemical thruster. The energies in the thruster are not high enough for equilibrium ionization, and the plumes are usually too collisional for nonequilibrium ionization to take place. However, anomalous ionization may be caused by the critical ionization velocity (CIV) effect (Newell, 1985).

Another mechanism for thruster contamination is due to the recoil of exhaust products back toward the spacecraft following collisions with atoms in the ambient environment or with slow-moving particles in the plume. For collisions with other exhaust products, only very light molecules in the plume colliding with heavy, slow-moving molecules would lead to recoil back toward the spacecraft.

The major mechanism for thruster contamination is slow-moving exhaust products in the boundary layer of the rocket nozzle, which recirculate back toward the spacecraft (Dettleff, 1991). The basic features of a plume expanding into a vacuum are shown in Figure 4.13. The flow is a continuum in the thruster with an isentropic core and a slower-moving boundary layer. The isentropic core undergoes a continuum expansion and then passes into the free-molecular-flow regime as the density drops. The boundary layer turns sharply around the nozzle lip while undergoing a continuum expansion. It then rapidly falls into a free molecular flow in the

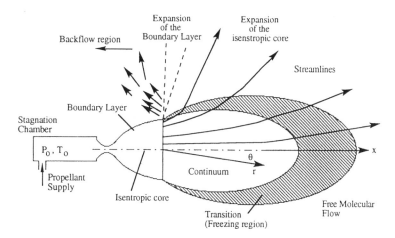

Figure 4.13. Flow Types in a Plume Expanding into Vacuum. Reprinted from *Progress in Aerospace Science*, vol. 28, Dettleff, Plume flow and plume impingement in space technology, pp. 1–71, Copyright 1991, Kind Permission from Elsevier Science Ltd., the Boulevard, Langford Lane, Kidlington, OX5 1GB, UK.

backflow regime. Since the dimensions of the flowfield of interest for contamination are much larger than the dimensions of the typical nozzle exit, the plume flow is modeled as emanating from a point source on the center line of the exit plane. Hence, the plume density far from the nozzle exit is given by

$$\rho(r, \Theta)/\rho^* = A_p(r^*/r)^2 f(\Theta), \tag{4.70}$$

for some constant A_p and angular function f and where ρ^* is the density at the nozzle throat and r^* is the radius at the throat. On the basis of isentropic flow theory as well as experiments, the function f for the isentropic core is given by

$$f(\Theta) = \cos^{\frac{2}{\gamma-1}}[(\pi/2)(\Theta/\Theta_{\lim})], \tag{4.71}$$

where $\Theta_{\lim}$ is the sum of the nozzle angle Θ_E and the Prandtl–Meyer deflection angle Θ_∞ (Vincenti and Kruger, 1965). The Prandtl–Meyer deflection angle is $\Theta_\infty = \nu_{PM}(M = \infty) - \nu_{PM}(M_E)$ where M_E is the Mach number at the nozzle exit and ν_{PM} is the Prandtl–Meyer function (Vincenti and Kruger, 1965). The expression in Eq. (4.71) is valid up to the stream line that emanates from the edge of the boundary layer at the nozzle exit. If the boundary layer at the nozzle exit of radius r_E has thickness δ, then the limiting stream line will exit at an angle Θ_o given by

$$\Theta_o = \Theta_{\lim}\left[1 - \frac{2}{\pi}\left(\frac{2\delta}{r_E}\right)^{(\gamma-1)/(\gamma+1)}\right]. \tag{4.72}$$

For angles $\Theta > \Theta_o$, the function f is modeled as (Simons, 1972)

$$f(\Theta) = f(\Theta_o)\exp[-\beta(\Theta - \Theta_o)], \tag{4.73}$$

where the quantity β is

$$\beta = A_p\left(\frac{\gamma+1}{\gamma-1}\right)^{1/2}\frac{2\bar{u}_{\lim}}{u_{\lim}}\left(\frac{r_E}{2\delta}\right)^{(\gamma-1)/(\gamma+1)}. \tag{4.74}$$

The limiting velocity $u_{\lim}$ is the maximum velocity in an isentropic flow given by

$$u_{\lim} = \sqrt{\frac{2\gamma}{2\gamma-1}R_g T_c}, \tag{4.75}$$

with R_g being the gas constant for the nozzle gas and T_c being the chamber temperature. The velocity $\bar{u}_{\lim}$ is the average velocity in the boundary layer and is usually taken as $\bar{u}_{\lim} = 0.75u_{\lim}$. The plume constant A_p follows from conservation of mass

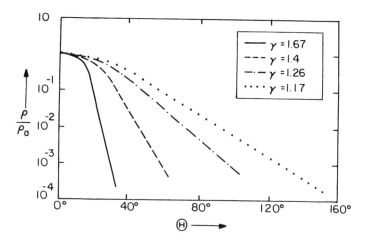

Figure 4.14. Angular Far-Field Density Distribution for a Conical Nozzle with an Area Ratio of 25 and Nozzle Exit Angle of 10 deg. Reprinted from *Progress in Aerospace Science*, vol. 28, Dettleff, Plume flow and plume impingement in space technology, pp. 1–71, Copyright 1991, Kind Permission from Elsevier Science Ltd., the Boulevard, Langford Lane, Kidlington, OX5 1GB, UK.

and is given by

$$A_p = \frac{u^*(2u_{\text{lim}})}{\int_0^{\Theta_{\text{lim}}} f(\Theta)\sin\Theta\,d\Theta}. \tag{4.76}$$

The angular far-field density distribution from this model is plotted in Figure 4.14 for several values of γ. The far-field density distribution matches the experimental data well for the plume regions that can be described by continuum flow. The onset of free molecular flow is described by a breakdown parameter (Bird, 1976), which for one-dimensional steady flow is

$$P = \frac{u}{\rho v}\left|\frac{d\rho}{dr}\right|, \tag{4.77}$$

where v is the collision frequency for molecules in the plume. A measure of the number of collisions that occur in the time that it takes for the gas to move through one density-gradient-length scale is $1/P$. Clearly, the flow is continuum if $P \ll 1$. If $\rho \sim 1/r^2$, $d\rho/dr \sim 1/r^3$, $nu \sim \rho$, and $u \sim$ constant, then $P \sim r$, so that there must be a radius at which continuum flow breaks down. Bird (1976), by using a direct-simulation Monte Carlo model found that the boundary is at $P \simeq 0.05$. For radial distances that are outside this region, the rarefied expansion flow is characterized by increasing free paths. The flow direction will still be radial but deflections will occur because of (rare) intermolecular collisions. Heavy molecules will tend to continue in a radial direction, whereas light molecules will undergo large deflections. As a consequence, heavy molecules will remain in the near-axis region, and light molecules will be found in the outer regions. This distribution of species could be of concern for some contamination situations.

4.3.3 The Shuttle Neutral Environment

Because the *Space Shuttle* is one of the major spacecraft used in LEO, there have been many measurements of its environment [see review by Green et al. (1985)]. The *Shuttle* has several means of depositing gas into the environment in addition to outgassing. It has a water FES and a mechanism for dumping large amounts of water. In addition, there is a vernier system of six rockets that provide 1.1×10^7 Nt of thrust and a reaction control system consisting of 38 thrusters, each of which can produce 3.8×10^8 Nt of thrust. All of these engines are bipropellant engines, which burn monomethylhydrazine (MMH) in nitrogen tetroxide. The exhaust products are mainly H_2O and N_2 with lesser amounts of H_2, CO_2, CO, O_2, and unburnt fuel.

Species that have been measured in the payload bay include He, H_2O, N_2, CO_2, and NO, among others. Most of the contamination in the bay is H_2O with pressures measured as high as 10^{-4} torr. By comparison, the ambient pressure is approximately 10^{-7} torr. The higher pressures are closely correlated with ram activities and thruster firings (Wulf and von Zahn, 1986). Normally, the pressure in the bay is approximately one order of magnitude higher than the ambient pressure.

4.4 Erosion by Atomic Oxygen

The degradation of spacecraft surfaces because of erosion or recession is a newly discovered phenomenon associated with LEO. This environmental effect was first observed during early Shuttle flights (Peters, Linton, and Miller, 1983; Gull et al., 1985; Peters, Gregory, and Swann, 1986). Surface erosion was seen on ram or forward-facing surfaces of several types of materials. Because the major component of the environment between 200 and 800 km is atomic oxygen, a strong oxidizing agent, and because the oxygen flux, which tends to dominate in this region, is approximately 10^{19} atoms/m^2-s, it is assumed that this is the principal source of erosion. In support of this, atomic oxygen, because of the high orbital velocity (≈ 7.8 km/s) of a spacecraft at these altitudes, is impacting the surface with an energy in excess of 5 eV – well above the energy necessary for a chemical reaction with many materials. Oxygen as the source implies that organic materials will be the most sensitive to this effect and that the amount of surface change should be proportional to the atomic oxygen fluence. This fluence is dependent upon attitude, altitude, exposure time, and solar activity, factors all borne out by subsequent in-situ experiments.

Erosion due to atomic oxygen is caused by the formation of volatile oxides (e.g., CO) or oxides that do not adhere to the surface (e.g., silver oxides) (Tennyson, 1993). Erosion is also enhanced by UV radiation, which can break molecular bonds and leave molecules that are easily oxidized by the atomic oxygen. Note that atomic oxygen does not always lead to erosion; it may also give rise to a stable oxide that stays on the surface. For example, silicon exposed to atomic oxygen develops a stable silicon dioxide coating.

A possible mechanism for surface erosion of graphite can be better understood by assuming that the graphite surface is directly exposed to the atomic oxygen flux (Tennyson, 1993). The strength of the carbon–carbon bond is 7.4 eV. If the surface is struck by an oxygen atom (whose energy can be up to 7 eV), then there is some finite probability that CO will form (bond strength of 13.1 eV). This is illustrated in Figure 4.15. Once such a bond is formed, if a second oxygen atom with energy greater than 1.7 eV strikes the surface, it can transfer sufficient energy (11.4 eV

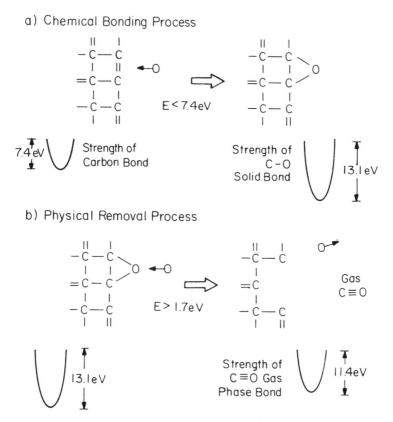

Figure 4.15. Atomic Oxygen Erosion Process for a Graphite Surface. From DeWitt et al. (1993). Reprinted by Permission of Kluwer Academic Publishers.

bond strength) to the surface to allow the CO to escape in the gas phase. This leaves a vacant site on the original graphite surface.

To date, most atomic oxygen erosion rates have been obtained from experiments on the Shuttle flights STS-5, STS-8, ground-based experiments, and the *LDEF*. One common material widely used on spacecraft surfaces, polyimide Kapton, has been found to be quite vulnerable to oxidation by the LEO atomic oxygen. For this type of Kapton, the volume of material oxidized per impacting oxygen atom is approximately 3×10^{-30} m^3/atom (Leger and Visentine, 1986). This yield has been found experimentally to be a weak function of impacting energy for energies of a few electron volts. The energy dependence for Kapton has been determined to scale as $E_i^{0.68}$ with E_i being the impacting energy. Hence, for Kapton on a surface with unit normal $\vec{n}$, the surface recession can be written as

$$\delta_r = RF_t, \tag{4.78}$$

where the fluence F_t is given by

$$F_t = \int_0^t dt \int d^3 v \vec{v} \cdot \vec{n} f_O, \tag{4.79}$$

and the reaction efficiency $R = R_c(E_i/E_{i0})^{0.68}$ with $E_{i0} = 5$ eV. This type of energy dependence is not known for many materials, and so Eq. (4.78) generally is used with $R = R_c$ taken as an empirically determined constant.

The STS-5 and STS-8 flights each provided an exposure of 41 hours to ram conditions. However, because of altitude, attitude, and solar activity differences, their total fluences were different: 10^{24} atoms/m^2 on STS-5 and 3.5×10^{24} atoms/m^2 on STS-8. For STS-5, the velocity vector rotated about the surface normal once every orbit, whereas, on STS-8, it was normal to the surface. Because the exposures resulted in loss of material as high as 12 μm of surface thickness, measurements in the change in mass of the samples provided an assessment of erosion rates.

Results from the *Shuttle* flights are listed in Table 4.4. Data are typically reported in terms of the reactivity (i.e., the thickness of material loss normalized to total oxygen fluence as estimated from atmospheric models, spacecraft velocity, and exposure history). Some samples contained fillers that were stable to atomic oxygen and shadowed or protected the underlying material from mass loss. As illustrated in Figure 4.16 for Kapton (Leger and Visentine, 1986), these reaction rates can be used for an assessment of the effects of atomic oxygen erosion on spacecraft surfaces. The following are general observations from Leger and Visentine (1986), Maag (1988), and Koontz, Albyn, and Leger (1991).

(1) Unfilled organic materials containing C, H, O, N, and S all react with approximately the same efficiency ($2-4 \times 10^{-30}$ m^3/atom).

Table 4.4. *Reactivity for materials exposed to LEO atomic oxygen*

Material	R_c (m^3/atom $\times$ 10^{30})
Kapton	3
Mylar	3.4
Tedlar	3.2
Polyethylene	3.7
Teflon FEP and FE	<0.1
Carbon	1.2
Polystyrene	1.7
Polyimide	3.3
Platinum	0
Copper	0.05

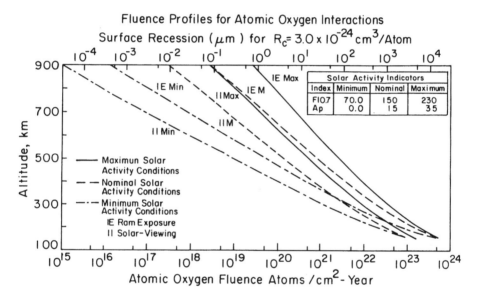

Figure 4.16. Fluence Profiles for Atomic Oxygen Interactions with Kapton. From Leger and Visentine (1986). Copyright © 1986 AIAA. Reprinted with Permission.

(2) Prefluorinated carbon-based polymers and silicones have lower reaction efficiencies than organics by a factor of 10 or more.

(3) Filled or composite materials have reaction efficiencies that are strongly dependent on the characteristics of the fillers.

(4) Metals, except silver and osmium, do not show macroscopic changes. Microscopic changes have been observed and should be investigated for systems very sensitive to surface properties. Silver and osmium react rapidly and are generally considered unacceptable for use in uncoated applications.

(5) Magnesium fluoride and oxides in various forms show good stability.

Table 4.5. *Flux and fluence requirements for space station for a*
15-year exposure

Surface orientation	Flux (atoms/m^2-s)	Fluence (atoms/m^2)
Ram	1.54×10^{18}	7.27×10^{26}
Solar facing	6.76×10^{17}	1.60×10^{26}
Antisolar facing	1.07×10^{18}	2.52×10^{26}

For spacecraft such as the *space station* (Leger and Visentine, 1986) which may be compromised by atomic oxygen erosion of its surface, there are two possible mitigation techniques: to select materials with low reaction rates, and to provide protective coatings on materials. Unfortunately, because many materials used in space are chosen for specific properties such as resistivity and thermal properties, and because it may not be possible to modify the orbit or change the mission time, most of the effort to mitigate atomic oxygen effects has focused on providing protective coatings. The need for this is illustrated by the following calculation. For the *space station* over a 15-year period, the atomic oxygen flux and fluence levels are given in Table 4.5 for solar conditions two standard deviations above a nominal *space station* mission (Banks et al., 1990). For solar arrays that have an unprotected exposed Kapton substrate, the reactivity data in Table 4.4 indicate that the solar and antisolar surfaces would lose 1.2 mm of Kapton in 15 years. This is much larger than the optimum desired thickness of 0.05 mm. Therefore, the unprotected Kapton would be completely eroded from the station and compromise the viability of the solar arrays. One protective coating that has been considered for Kapton is SiO_2 deposited by sputtering. Other types of coatings include Al_2O_3 as well as indium tin oxide (ITO).

An important issue that arises with protective coatings is the effect of defects in the coating. For example, SiO_2 deposited by high-frequency magnetron sputtering has a defect density of up to 6×10^8 defects/m^2. At a defect, the underlying Kapton is exposed to atomic oxygen, which can enter over a range of angles through the defect hole. This occurs because the thermal distribution of oxygen leads to a small dispersion in the incoming velocity vectors. Also, the ram flux will sweep across the surface at a varying angle because the protected surface is not always oriented toward the ram. Both effects lead to an undercutting of the defect site, as illustrated in Figure 4.17. Once several defect sites are sufficiently undercut that they start to interact, the overlying coating can tear and expose more of the Kapton to attack. This trend is illustrated in Figure 4.18 where the erosion rate for protected Kapton is given as a function of fluence for a plasma asher (Banks et al., 1990). If there were no undercutting and subsequent tearing, after the Kapton had been removed

Duration of Atomic
Oxygen Exposure   Side View   Front View

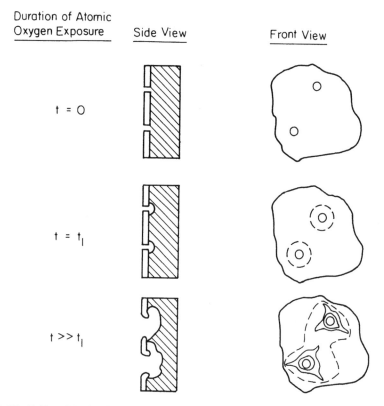

t = 0

t = t₁

t >> t₁

Figure 4.17. Failure Mode of Protective Coatings as a Result of Atomic Oxygen Undercutting. From Banks et al. (1990). Used by Permission of the Author.

below the hole, the mass rate loss would saturate as a function of fluence. The fact that it keeps on increasing is due to the undercutting and tearing. Nevertheless, the SiO_2-protected Kapton is clearly more resistant to atomic oxygen attack than the unprotected Kapton.

4.5 Glow

Vehicle glow is another phenomenon detected in the previous decade at low altitudes (below 800 km). It appears as a dim background light source near or on the surface of spacecraft. Although suspected on several early rocket flights, vehicle glow went essentially undetected until the late 1970s. It has since been reported in numerous papers [see reviews by Green et al. (1985) and Garrett, Chutjian, and Gabriel, (1988)]. Glow, because it may potentially affect measurements in the IR, UV, and visible spectral bands of importance to many optical sensor systems, can be a particularly serious contamination issue. In Figure 4.19, the glow surface brightness from several *Shuttle* missions has been compared with Atmospheric *Explorer*

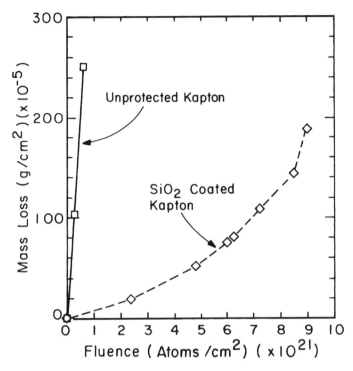

Figure 4.18. Mass Loss of SiO_2-Protected Kapton as a Function of Fluence in an RF Plasma Asher. From Banks et al. (1990). Used by Permission of the Author.

(AE) glow measurements between 4,278 Å and 7,320 Å. Although the comparison is somewhat subjective considering the possible geometric, thermal [see Swenson et al., (1986)], and temporal (i.e., changes in the ambient neutral density with solar activity) differences between the observations, the agreement is good, with the STS 41-G discrepancy being ascribed to a thermal effect (the surfaces were warm). The figure clearly shows the most pronounced feature of glow: its variation with altitude (believed to be due primarily to the exponential falloff of the atomic oxygen density with altitude). Other features of the glow are summarized below.

Although they may be due to one phenomenon, there are apparently several different types of glow. These can be crudely defined as satellite glow and *Shuttle* glow. *Shuttle* glow is further divided into ram glow, thruster glow, and cloud glow. General features of these types of glow are the following:

(1) Most glow is maximized on surfaces facing into velocity vector or ram direction (note that cloud glows may have been observed in the wake).

(2) The intensity of glow decays with altitude. In the case of AE-C, it decreased exponentially with altitude; the scale height was roughly 35 km, consistent with atomic oxygen at a temperature of 600 K. As illustrated in Figure 4.19, there may be strong similarities in how the intensity scales with height between the *Shuttle* and satellites.

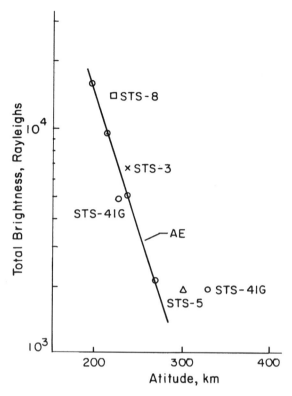

Figure 4.19. Comparison of the Satellite (AE) and Shuttle Surface Glows as a Function of Altitude. [From Bareiss et al. (1986)].

Small bodies in space appear to have a characteristic glow associated with them. As yet, it is not clear whether this glow, called satellite glow, is identical to the better-known *Shuttle* ram glow. Its variation as a function of altitude for the AE spacecraft is plotted in Figure 4.19. The characteristics of satellite glow can be summarized as follows:

(1) The glow increases in brightness toward the red.
(2) The scale length of glow appears to be 1 to 10 m.
(3) The angular variation with respect to the ram direction is proportional to $\cos^3 \theta$.
(4) OH (possible Meinel bands) is the proposed source of satellite glow.

The best-known example of glow is that observed on the ram surfaces of the Shuttle. This form of glow is designated as *Shuttle* glow-ram. A typical picture, showing the variation with ram angle, is presented in Figure 4.20. The key features of this glow are the following:

(1) The glow is emitted over a continuum, rather than in discrete visible lines, extending throughout the visible, and having a peak intensity at 6,800 Å.

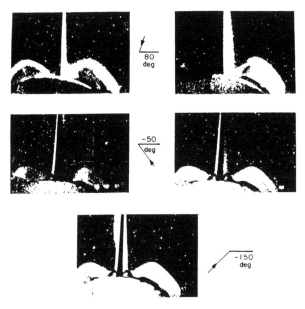

Figure 4.20. The Appearance of Glow on Different Parts of the *Shuttle* During the Roll Experiment on STS-5. Arrows Indicate the Velocity Vector. From Garrett et al. (1988). Coyright © 1988 AIAA. Reprinted with Permission.

(2) The scale length of the glowing layer in the ram direction is estimated at 20 cm above large flat surfaces on the *Shuttle* (perhaps only 6 cm above the RMS), which is consistent with an effective radiative lifetime of emitting molecule(s) of about 0.6–0.7 ms for a mean emitter velocity of 0.3 km/s.

(3) The intensity of glow varies from material to material with black chemglaze (carbon-filled, urethane-based paint) and Zn_3O_2 (overcoated with Si) being the brightest and polyethylene being the least bright. The scale length, however, is similar for each sample.

(4) The angular variation with ram angle is closer to $\cos\theta$ than $\cos^3\theta$.

(5) There is an apparent strong exponential correlation between surface glow intensity and surface temperature, with intensity increasing with decreasing temperature. In support of this, the weak AE satellite glow relative to *Shuttle* glow is consistent with the respective differences in surface temperature.

This *Shuttle* glow can be divided into three regions: visible, far ultraviolet (FUV), and IR. One theory for the visible *Shuttle* glow is that it is due to emission from excited NO_2 produced on the *Shuttle* surface by a surface reaction called a Langmuir–Hinshelwood (L–H) process. The reaction is between O and NO on the surface (Viereck et al., 1991) and is written as

$$O_{fast} + NO_{ads} \rightarrow NO_2^*$$
$$NO_2^* \rightarrow NO_2 + h\nu. \tag{4.80}$$

The oxygen to form the NO_2 comes from the ambient atmosphere whereas the NO is adsorbed on the surface. One possible source for the NO is from thruster firings, although equilibrium calculations of thruster exhausts do not predict formation of NO. The other possible source is the gas-phase reaction

$$O + N_2 \rightarrow NO + N. \tag{4.81}$$

The NO would then migrate to the surface and be adsorbed. The FUV glow is due to excited molecular nitrogen emissions, whereas the IR emissions are due to excited H_2O. The H_2O becomes collisionally excited from the ramming atmospheric oxygen and then cascades to produce emission in the wavelength range of 2.5–14 μm.

A second type of glow observed on the *Shuttle* is that associated with thruster firings. Currently, there are few measurements of this glow. Because of the apparently large background enhancements it produces (values as high as 10^6 Rayleighs have been observed), it is potentially a major concern for large space structures that will need to be continuously firing attitude control thrusters. This *Shuttle* thruster glow can be characterized as follows:

(1) The glow is enhanced after firing of the Shuttle's attitude thrusters (note that thruster effluents primarily consist of H_2O).
(2) The thruster-induced glow, measured at 4 Å, was a continuum in the wavelength range 6,275–6,307 Å over which the measurement was made.

5

Plasma Interactions

5.1 Spacecraft–Plasma Interactions

The plasma environment that a spacecraft will encounter as a function of orbit is described in Section 3.3. Although the plasma environment is not necessarily the dominant environment in a particular case, it can nevertheless have a profound and destructive effect on a spacecraft or its payload. In particular, major plasma effects follow from the slow accumulation of charge on surfaces. This accumulation of charged particles from the surrounding space plasma on spacecraft surfaces, termed surface charging, produces electrostatic fields that extend from surfaces into space and can result in a number of adverse interactions:

- surface arc discharges that generate electromagnetic interference cause surface damage, induce currents in electronic systems, produce optical emissions, and enhance the local plasma density;
- enhanced contamination leading to changes in surface, thermal, and optical properties;
- a shift of the spacecraft electrical ground, leading to problems with detectors collecting charged particles from the environment; and
- coulomb forces on the spacecraft components and materials as well as modifications of the drag coefficient and electromagnetic torques on the spacecraft.

In addition to these concerns, there are some less obvious effects. Of particular concern to the manned spacecraft community, differential charge accumulation between two spacecraft that come into contact (the *Shuttle* and the station or an astronaut during extra-vehicular activity and the *Shuttle*) may result in damaging current flows between the spacecraft. These can cause arc discharges, electronic burnout, and other safety hazards. The continued flow of charge to and from the spacecraft surfaces constitutes a current that can lead to electrical power leakage from the spacecraft. (Note that in some circumstances this is not bad because this current, in principle, can be utilized as a source of power or as a source of thrust for the spacecraft.) When there is a large, artificially generated plasma around a spacecraft, it may substantially modify the spacecraft ground, discharge the space-craft surfaces, or interfere with the communication links to and from the spacecraft.

Even the acceleration of charged particles into the spacecraft surface can be of concern. The impact of the ions with the surface at high energies can lead to surface chemical reactions with associated degradation or to surface sputtering and contamination. Electron impact on surfaces can lead to enhanced electron emission as well as desorption of neutral gas from the surface.

The plasma wake of a spacecraft in LEO can make it easily detectable. In addition, the emission of plasma or the creation of plasma around a spacecraft can modify and enhance the vehicle's signature as seen by sensors in its far field. The plasma cloud around a spacecraft also may be a source of electromagnetic and electrostatic waves. Although these can interfere with onboard electrical systems, the interaction of these waves with the ambient environment also may reveal a great deal of information about the physics of the ambient medium.

As can be seen from this discussion, there is a rich set of possible interactions between the space plasma, either natural or artificial, and a spacecraft. To under-stand them, it is necessary first to discuss the general theory of charge and current collection by an isolated body. Building on this, interactions at GEO and higher-altitude orbits and then LEO and polar orbits are discussed. The reason for this division is that for GEO, the Debye length is so much larger than the spacecraft that the plasma effectively behaves as a collection of isolated charged particles rather than a set of coupled particles. This is not true at lower altitudes and makes the latter interactions more complex as well as harder to understand.

5.2 Spacecraft Surface Charging and Current Collection

A spacecraft acts like an isolated electrical probe in the space plasma. Like any electrical probe in a plasma (Chen, 1984), it will collect charge and adopt an electrostatic potential consistent with charge collection as required by Maxwell's equations. This collection of charge from the environment has been called "space-craft charging." Three distinct forms of spacecraft charging have been identified as potential sources of concern. First, there is vehicle charging in which the over-all satellite ground potential varies relative to the ambient space plasma potential (taken to be zero far from the spacecraft). This type of charging, first observed by DeForest (1972) at geosynchronous orbit, may reach tens of thousands of volts and significantly distort plasma observations, but, in general, does not pose a hazard to spacecraft operations. The second type of charging involves the deposition of high-energy ($E > 100$ keV) electrons in dielectrics or on isolated conducting surfaces inside the vehicle's outer surface. Electric fields in excess of 10^6 V/cm can be built up, leading to the breakdown of most common high-voltage insulators. This type of charging can be particularly destructive because it can occur on or near internal circuitry, which typically is poorly shielded. The third form of charging, differential surface charging, results from potential differences on the surface of a spacecraft of a few volts to several thousands of volts. Although typically of a lower potential difference than the other two forms, this type of charging is often more destructive because it leads to damaging surface arcing.

The intent of this section is to review charging of the vehicle as a whole and in terms of each of the three charging processes. Given the spacecraft designer's concern with arcing, however, emphasis is placed on estimating differential charging. The latter form of charging is, however, very dependent on three-dimensional effects (in particular, the details of the geometry of the vehicle) making generalizations difficult. Following a brief description of its effects, the sources of charging are discussed. Theoretical models then are reviewed to illustrate processes by which differential charging may occur.

Arcing, defined as the rapid (approximately nanosecond to microsecond) re-arrangement of charge by punchthrough (breakdown from dielectric to substrate), by flashover (propagating subsurface discharge), by blowoff (arc to surface), between surfaces, or between surfaces and space, is not well understood. Fortunately, the effects of arcing are somewhat better understood than the process itself. Specifically, the arc-caused pulse generates a transient in spacecraft electrical systems either by direct current injection or by induced currents due to the associated electromagnetic wave. In addition to the electrical damage, Balmain (1980) gives numerous examples of the micrometer-size holes and channels that are found in dielectric surfaces following the arc discharge. Nanevicz and Adamo (1980) have found additional, large-scale physical damage to solar cells, such as fracturing of the cover glass. Similar damage to circuit elements can result in burnout or breakage of individual components. Because this latter damage is often hard to observe, isolation of the discharge effects during ground testing can be difficult, adding to the problem.

Also of particular concern to the space physics community are the effects of differential spacecraft charging on plasma measurements. There are numerous ways that such charging can complicate the interpretation of low-energy plasma measurements. These can be characterized loosely as shifting of the spectra in energy, preferential focusing or exclusion of particles of a particular energy or direction, and contamination of measurements by secondaries, backscattered electrons, and photoelectrons. In the simplest example of charging effects on measurements, a potential difference between the vehicle and space can raise or lower the energy of the incoming particles. In the case of differential flux measurements such as by electrostatic analyzers, the shift in energy is easily detected in the attracted species – the ions. The most complex variations, however, are caused by differential charging, where particle focusing or exclusion can occur. Differential charging of surfaces produces electric fields that can focus or defocus incoming particles at specific energies. Just as the geometry of the vehicle can shadow the field of view of a detector, the potential gradients near the field of view of a detector can distort the particle trajectories.

Contaminant ions, from thrusters (ionic or chemical) or outgassing of satellite materials, can be trapped within the satellite sheath and preferentially deposited on negatively charged spacecraft surfaces. It has been estimated (Jemiola, 1980) that as much as 50 Å of material can be deposited on charged optical surfaces in as little as 100 days. Nanevicz and Adamo (1980) found that the heating rate of sensors on a geosynchronous satellite rose with geomagnetic activity. This is believed to be due to increased contaminant deposition during periods of geomagnetic activity and, therefore, increased charging. Such deposition also may alter secondary emission and photoelectron properties.

There are two major sources of differential charging: mechanical and charging-induced. Although both are intimately related and typically occur together, mechanical sources are defined as those caused by processes independent of the charging. Ambient flux anisotropies and photon shadowing are examples. Charging-induced sources are those caused by the charging process itself. Examples are deviations in particle trajectories because of charged surfaces and space charge as well as differences in the dielectric properties of different parts of a surface.

5.2.1 Current Sources to a Spacecraft

5.2.1.1 Current from the Ambient Plasma

The primary natural source for potentials of 10 kV or higher on spacecraft surfaces is the ambient space plasma. Although the space plasma is seldom representable in terms of a single temperature and density (see Section 3.3.4), the Maxwellian distribution function is a useful starting point for describing the ambient plasma conditions that generate these large potentials. Assuming the distribution function is that for an isotropic Maxwellian plasma, the current flux to a surface at rest in the plasma (the situation that prevails at GEO) and at the plasma potential (i.e., no electric field between the surface and the far-field plasma) is from Eq. (2.29):

$$j_a = -e(n_i \bar{c}_i / 4 - n_e \bar{c}_e / 4). \tag{5.1}$$

The notation adopted in this equation and in subsequent equations is that a current (or flux) of positive particles *to* the surface is defined as negative, and the sign of a given current is always explicitly written in any equation where that current appears. The net current is the difference between the incoming and outgoing currents. Typical values of this current (or flux) at GEO are listed in Tables 3.11

and 3.12. The median current from these tables is of the order of 1–5 μA/m^2. It can be seen from a comparison of Tables 3.11 and 3.12 that the dominant contribution to the current density at zero potential comes from the electrons. This is a consequence of the electrons' much smaller mass relative to that of the ions, which makes them much more mobile.

In-situ measurements have shown that the ambient plasma flux can be quite anisotropic with large pitch angle variations. The result is that j_a can vary over the surface of the satellite in a complex, three-dimensional fashion, leading directly, for electrically isolated surfaces, to differential charging. Such variations were clearly evident on the *SCATHA* spacecraft. When this effect and those associated with the spacecraft potential, sheath anisotropies, and deviations of the ambient plasma from a Maxwellian distribution are considered, the simple distribution model inherent in Eq. (5.1) is no longer valid. Even so, Eq. (5.1) and modifications of it are accurate for many practical purposes and are used often in the discussion that follows.

For LEO and polar orbits, the motion of the spacecraft is mesothermal (Section 2.4.3). Therefore, for a surface at the plasma potential, the electron collection will be as in Eq. (5.1), but the ion collection will be dominated by the ram flux (i.e., the flux that directly intercepts the body). Hence, the current flux to a surface oriented in the ram direction at zero potential is

$$j_a = -e(n_i V_o - n_e \bar{c}_e/4). \tag{5.2}$$

This current density can be evaluated using the numbers in Table 3.8 for the *space station* orbit. Once again, the electron current dominates at zero potential and typically gives current densities of 1–10 mA/m^2.

5.2.1.2 *Photoelectric Currents*

Many of the materials used on spacecraft surfaces emit photoelectrons when exposed to the UV component of the solar flux. Often this can be a significant contributor to the current to the spacecraft. The photoelectron current from a surface is a function of satellite material, solar flux, solar incidence angle, and satellite potential [see review by Lucas (1973)]. Figure 5.1, adapted from Grard (1973), is a plot of the two functions necessary to describe the photoelectron current. These are the solar flux as a function of energy, $S(E)$, and the electron yield per photon, $W(E)$. The product of these functions is shown in Figure 5.1. For a photoemitting surface at zero or negative potential, the electrons will all escape and the photoelectric current density, therefore, is given by

$$j_{\mathrm{ph}_0} = -\int_0^\infty W(E)S(E)\,dE. \tag{5.3}$$

Table 5.1. *Photoelectron emission characteristics*

Material	Work function (eV)	$|j_{ph_0}|$ ($\mu A/m^2$)
Aluminum oxide	3.9	42
Indium oxide	4.8	30
Gold	4.8	29
Stainless steel	4.4	20
Aguadag	4.6	18
Lithium fluoride on gold	4.4	15
Vitreous carbon	4.8	13
Graphite	4.7	4

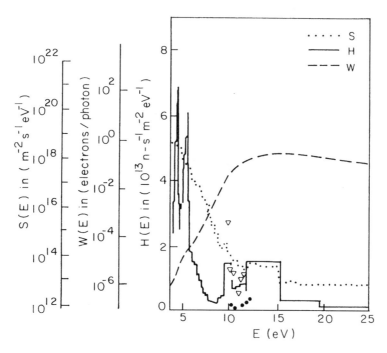

Figure 5.1. Composite Plot of the Electron Yield per Photon $W(E)$, the Solar Flux $S(E)$, and Their Product, the Total Photoelectron Yield, as Functions of Energy, E, for Aluminum Oxide.

The total possible photoelectron current density at 1 AU for a variety of materials, j_{ph_0}, is tabulated in Table 5.1.

The photoelectron current is often the dominant current for high orbits (compare typical values from Table 5.1 of tens of microamperes per square meter to ambient current fluxes of 1–5 $\mu A/m^2$). This is a major determinant of three-dimensional effects because it is, by its nature, anisotropic – occurring only on the sunlit surfaces of the vehicle. In this regard, illuminated cavities and raised edges (which can shadow a large surface at grazing incidence) can cause serious three-dimensional charging effects. This was observed on ATS-5 where the plasma detector that produced the ambient plasma data was recessed inside a cylindrical well. The surface of the well was nonconducting, and, as sunlight struck the surface, charged and discharged with a modulation at the spin period of the vehicle.

If the spacecraft is positively charged, the ambient photoelectron current is attracted to the surface. Because this return current is a function of potential and geometry, its accurate calculation requires that the energy spectrum of the emitted electrons for an incident photon at a given wavelength be known. Both laboratory and space experiments indicate that photoelectrons are emitted isotropically with a Maxwellian energy distribution (Whipple, 1981). The mean energy of the distribution is 1–2 eV with very little dependence on the spectrum of the incident photons (except that they have to exceed the photoemission threshold). Grard (1973) has calculated the photoelectron return current for various materials and positively biased probe geometries. Variations in material across the spacecraft surface, as illustrated by these calculations, can lead to differences of factors of 10 in net photoelectron emission for adjacent surfaces. Such large variations can enhance differential charging effects.

5.2.1.3 Backscattered and Secondary Electrons

When an electron impacts a surface, it is reflected or absorbed into the material (Whipple, 1981). If it is absorbed, it may collide with atoms in the material and eventually reverse direction and backscatter out of the material. The electrons that do not backscatter lose energy in the material. Some of this energy can go into exciting other electrons, which may then escape from the material. This process is called secondary emission. The three processes are treated separately. The process of reflection is only significant at very low energies. For an electron striking a surface with almost zero primary energy, the reflection coefficient is of the order of 0.05 and decreases with increasing energy. In terms of energy, backscattered electrons are distinguished from secondary electrons by the energy of the emitted electrons. Backscattered electrons leave the material with an energy slightly lower than that of the primary electrons. Secondary electrons are emitted with a characteristic energy spectrum and leave the material with energies of, at most, a few eV. Typically, their energy spectrum is assumed to be a Maxwellian energy distribution with a mean energy of 2 eV.

Ions that impact a surface also may reflect and give rise to secondary electron emission. The flux of secondary electrons due to either electron or ion impact can exceed the incident fluxes under some circumstances and has proven to be an important contributor to spacecraft charging control. Obviously, for different materials, differing electron emission properties can cause three-dimensional variations. Surprisingly, however, the low emission energy of secondary electrons allows relatively small three-dimensional effects to suppress these fluxes. For reference, backscattered electrons typically have roughly half the original energy of the original electrons and about 20–30 percent of the original flux. They can be thought of as primarily reducing the incident flux by a fixed constant. Secondary electrons due to ions, because the ions are of opposite sign to the electrons, peak as the surface potential goes negative and approach perhaps 30–40 percent of the incident flux. For positive potentials, because the electron energy is typically only a few electron volts, the secondary flux can be considered to be zero. The secondary electrons due to electron impact are very complex in their effects as they go through an emission maximum (actually exceeding the number of incident electrons for many materials) near 200 to 1,000 eV. This is the level at which most geosynchronous spacecraft are typically charged. Their low emission energy (again, a few electron volts) means that even a low positive potential will suppress their flux. For incident fluxes of electrons with energies below 2 keV, spacecraft charging can be effectively prevented. Thus any process that can suppress secondary emissions can lead to large swings in spacecraft surface potential over the affected surface. The so-called snapover phenomenon for high-voltage solar arrays is believed to be due to this effect (see Section 5.2.3).

The yield curve for the emission of secondary electrons as a function of impact energy has been found by experiment to have a universal shape, as shown in Figure 5.2. This can be explained simply by noting that, for very small impact energies, the yield of secondary electrons has to be very small because there is not enough energy to displace the electrons. For very large energies, the primary electrons do not spend enough time in the vicinity of each atom to displace electrons,

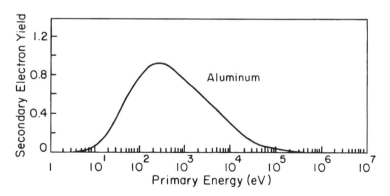

Figure 5.2. Secondary Electron Yield δ_e Due to Incident Electrons of Energy E Impacting Normally on Aluminum. From Garrett and Pike (1980). Copyright © 1980 AIAA. Reprinted with Permission.

and by the time the primary electron has slowed down, it is too deep in the material for the secondaries to escape. For secondary yield curves that exceed unity, there are obviously two unity crossings. The first unity crossing is defined as the lowest energy at which the yield is unity. The second unity crossing is defined as the highest energy at which the yield is unity. In general, the yield is also a function of the angle at which the electron impacts the material, increasing as the angle increases away from normal. This is because secondaries are produced in a surface layer, and glancing electrons spend more time near the surface creating secondaries. There are several models of the secondary yield curve. One in common use (Sternglass, 1954; Chaky, Nonnast, and Enoch, 1981) is

$$\delta_e(E, \theta) = \delta_{e_{max}} \frac{E}{E_{max}} \exp\left(2 - 2\sqrt{\frac{E}{E_{max}}}\right) \exp[2(1 - \cos\theta)], \quad (5.4)$$

where the primary electron impacts with energy E at an angle from the surface normal of θ. The yield function is completely characterized by the maximum yield $\delta_{e_{max}}$ which occurs at the primary energy E_{max}. Another expression that is used in the computational modeling program NASCAP (Katz et al., 1977; Whipple, 1981) is

$$\delta_e(E, \theta) = \frac{1.114\delta_{e_{max}}}{\cos\theta} \left(\frac{E_{max}}{E}\right)^{0.35} \left\{1 - \exp\left[-2.28\cos\theta\left(\frac{E_{max}}{E}\right)^{1.35}\right]\right\}.$$

$$(5.5)$$

This is also characterized by the maximum yield at a specific energy. The importance of correctly modeling the secondary electron yield is especially important at high energies because the yield has such a strong effect on the emitted electron current from a surface (Katz et al., 1986). The maximum yield and the energy at maximum yield have been measured for many spacecraft materials. They are incorporated into material databases, which are a critical component of spacecraft charging codes (Katz et al., 1977). Typical values are given in Table 5.2.

Once the yield function is known, the emitted current density due to secondary electrons for a negatively biased surface becomes

$$j_{se} = -e\frac{2\pi}{m_e^2} \int_0^\infty dE^* \int_0^\infty dE \int_0^\pi \sin\theta d\theta g(E^*, E)\delta_e(E, \theta)f(E), \quad (5.6)$$

where $g(E^*, E)$ is the normalized emission spectrum of secondary electrons at energy E^* due to incident electrons of energy E, and $f(E)$ is the energy distribution of the incident electrons. The function g is often approximately independent of E

Table 5.2. *Representative values for maximum yield and energy at maximum yield for secondary electron emissions by primary electrons*

Material	$\delta_{e_{max}}$	E_{max} eV
Aluminum	0.97	300
Aluminum oxide	1.5–1.9	350–1,300
Magnesium oxide	4.0	400
Silicon dioxide	2.4	400
Teflon	3	300
Kapton	2.1	150
Magnesium	0.92	250

and is usually modeled as a Maxwellian with a mean energy of $E_{sec} = 2$ eV. The incident electron current is given by

$$j_e = e\frac{2\pi}{m_e^2} \int_0^\infty dE \int_0^\pi \sin\theta d\theta f(E). \tag{5.7}$$

A comparison of Eqs. (5.6) and (5.7) shows that for materials where the maximum yield is substantially larger than unity, the secondary outgoing current density can exceed the incident current if the mean energy of the impacting electrons is near the energy for maximum yield.

For backscattered electrons, the outgoing current density for a negatively biased surface is given by

$$J_{be} = e\frac{2\pi}{m_e^2} \int_0^\infty dE^* \int_{E^*}^\infty dE \int_0^\pi \sin\theta d\theta B(E^*, E)\cos\theta f(E), \tag{5.8}$$

where $B(E^*, E) = G(E^*/E)/E$ and $G(x)$ is the percent of electrons backscattered at the fraction x of the incident energy E. A plot of $G(x)$ is given in Figure 5.3 for normal incidence of electrons on aluminum. The value of G is typically less than unity.

As noted previously, under ion bombardment, surfaces may emit secondary electrons (the probability of ion backscattering is very low and is usually ignored). Secondary electron emission yield due to ions can be approximated (Whipple, 1981) by

$$\delta_i(E, \theta) = \frac{2\delta_{i_{max}}\sqrt{E/E_{max}}}{1 + E/E_{max}} \sec\theta. \tag{5.9}$$

This equation has been found to fit the data for a number of metals under proton impact. In particular, for protons on aluminum, $\delta_{i_{max}} = 4.2$ at $E_{max} = 90$ keV.

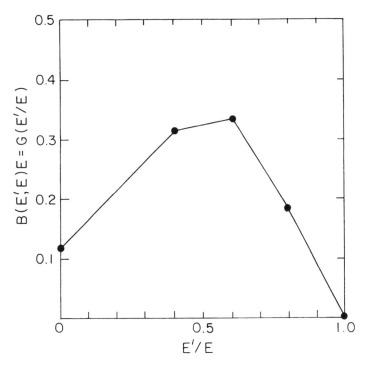

Figure 5.3. Percent of Electrons Scattered as a Function of the Fractional Energy Relative to the Incident Energy for Electrons on Aluminum. From Garrett and Pike (1980). Copyright © 1980 AIAA. Reprinted with Permission.

Typically, secondary electron yield under ion impact is only substantially larger than unity for incident impact energies above 10 keV. Once a yield function has been defined for a material, the outgoing secondary electron current can be defined as in Eq. (5.6), where the outgoing distribution function of the secondaries is represented by an isotropic Maxwellian with a mean energy of a few electron volts.

5.2.1.4 Effect of Magnetic Fields on Current Collection

For spacecraft in LEO, the magnetic field is sufficiently strong that the anisotropies introduced by the field must be considered in the collection of charge. The magnetic-field-induced anisotropies fall into two classes: motional electric fields or flux anisotropies. The first is the result of a spacecraft moving across a magnetic field. The spacecraft will see a motional electric field imposed on it given by

$$\vec{E}_m = \vec{V}_o \times \vec{B} \tag{5.10}$$

in MKS units. This motional field arises from the transformation of Maxwell's equations under translation and is fundamentally a consequence of the relativistic requirement that light always move at the same speed. The consequences of this motional field are that the potential of the spacecraft does not have a unique reference in the plasma. Instead, the potential difference between a spacecraft and the plasma, which drives the current collection, varies with position on the body even for a conducting body. This motional field is highest for PEO and LEO and is of the order of 0.25 V/m. Given the space station's potentially large size, the induced electric field can be significant for the *space station* (the potential difference from one end to the other may be as large as 26 V). The $\vec{V}_0 \times \vec{B}$ field is also the fundamental driver for an electrodynamic tether, where it can approach 5,000 V for a 20-km tether. It is also an issue for satellites in LEO with electric-field experiments that employ long (10 m or longer) antennas or booms. The local fields and current flows in the vicinity of the vehicle will be distorted by this effect.

In addition to the motional electric field, the magnetic field also induces aniso-tropies in the particle fluxes. Ambient fluxes, secondaries, beam fluxes, and charged-particle wakes are all controlled to a greater or lesser extent by the magnetic field. The key parameter in determining the influence of the magnetic field on the current collection is the magnetization parameter (see Section 2.4.1). For all but the largest spacecraft in LEO, the ion magnetization parameter is $M_i \gg 1$, so that the ion motion is ballistic as far as a charge collection is concerned. However, the electron parameter is $M_e \ll 1$, so that the electron motion toward the spacecraft as seen from the spacecraft is the E-cross-B drift under the motional electric field along with tight gyrations around a field line and free flow along the magnetic-field direction. Note that the E-cross-B drift evaluated on the motional field is

$$\vec{v}_d = \frac{\vec{E}_\perp \times \vec{B}}{B^2} = \frac{(\vec{V}_0 \times \vec{B}) \times \vec{B}}{B^2} = -\vec{V}_0.$$

Hence, in the spacecraft frame, the electrons are seen to flow toward the spacecraft with the velocity of the spacecraft, but, in a frame fixed to the Earth, the electrons would be seen as gyrating round a fixed-field line. This phenomenon is known as magnetic trapping. Specifically, if a charged particle is released from the spacecraft with a low velocity relative to the spacecraft or is created with low energy relative to the spacecraft, in the Earth's frame it would gyrate about the field line on which it was created. In the spacecraft frame, however, it would appear to be swept away at velocity V_0.

The anisotropy induced by the magnetic field means that the spacecraft can easily collect electrons from the direction $-\vec{V}_0$ and from the direction $\vec{B}$ but much less easily from the direction $\vec{V}_0 \times \vec{B}$. Parker and Murphy (1967) analyzed the effects of these magnetic-field-induced anisotropies on spacecraft charging and found that the electron flux can be reduced by as much as a factor of two on some surfaces. Gyrations around the field can also strongly affect the escape of secondary electrons from a spacecraft surface. If the magnetic field is parallel to the surface, the emitted electrons will gyrate around the field and return to the surface – they will not leave the surface and will not contribute to the outgoing current flow.

The current collection by probes and electrodes in space magnetoplasmas was reviewed by Laframboise and Sonmor (1993). They noted that current collection by a body in a magnetic field is most readily understood for a stationary collisionless system. However, real spacecraft in LEO are continually in motion and are surrounded by an environment that is not only collisional but is also very turbulent. Therefore the actual current collection by a real spacecraft may differ substantially from the predictions of simple models. This is addressed in Section 5.3.2.

5.2.1.5 Artificial Current and Charge Sources

Artificial mechanisms that affect spacecraft charging are numerous and include electron and ion beams, exposed, high-potential surfaces such as the junctions between solar cells (Stevens, 1980), highly biased electrodes (Katz et al., 1989), and plasma contactors (Hastings, 1987b; Dobrownoly and Melchioni, 1993). Plasma beams have been actively exploited both as probes of the satellite sheath and as a means of controlling the spacecraft potential (Goldstein and DeForest, 1976; Purvis and Bartlett, 1980; Dobrownoly and Melchioni, 1993). Kilovolt potentials and currents between milliamperes and amperes are typical of these systems. In the TSS-1 experiment on the *Shuttle*, the electron gun was capable of emitting 0.7 A at 5 kV (Dobrownoly and Melchioni, 1993) whereas for the neutralization experiment, reported by Sasaki et al. (1987), the electron gun emitted 300 mA at 4.9 kV.

5.2.2 General Probe Theory

The fundamental equation for charging theory is that of current balance. This equation states that, for a surface element on a spacecraft with charge density σ and area A, the rate at which the charge on the surface changes as a function of time must be given by the difference between all the currents to the surface element and all the currents away from the surface element. That is,

$$\frac{d\sigma}{dt} A = I_{\text{from surface}} - I_{\text{to surface}} = I_{\text{net}}. \tag{5.11}$$

This is illustrated in Figure 5.4, which shows all of the sources of current to an isolated conductor or a dielectric layer on a spacecraft. Eq. (5.11) follows directly from conservation of charge and the definition of current. This can be seen from Maxwell's equations by the following argument for a spherical conducting body.

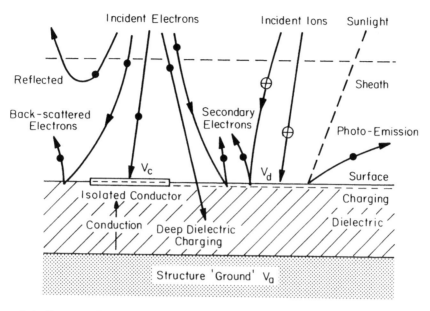

Figure 5.4. Currents that Control Charging on a Spacecraft Surface-Isolated Conductor or a Dielectric Layer. From DeWitt et al. (1993). Reprinted by Permission of Kluwer Academic Press.

From the divergence of the Poisson equation and Ampere's law it can be seen that

$$\frac{\partial \rho}{\partial t} + \nabla \cdot \vec{j} = 0, \tag{5.12}$$

where ρ is the charge density. If this equation is integrated over all space outside the sphere and the divergence theorem is used, then

$$\frac{\partial Q}{\partial t} = I_{net}, \tag{5.13}$$

where Q is the charge stored on the sphere, and

$$I_{net} = \int_S \vec{j} \cdot \vec{n} \, dS, \tag{5.14}$$

and the unit normal vector points out from the surface S. Eq. (5.11) or Eq. (5.13) is known as the charging equation.

For steady state or at equilibrium, the charging equation states that the net current to a surface must be zero – otherwise charge of one sign or the other will accumulate. Therefore, the equation expressing this current balance for a given surface in an equilibrium charging situation is, in terms of the currents,

$$I_e(V_s) - [I_i(V_s) + I_{se}(V_s) + I_{si}(V_s) + I_{be}(V_s) + I_{ph}(V_s) + I_b(V_s) + I_s(V_s)]$$
$$= I_{net} = 0, \tag{5.15}$$

where

V_s = satellite surface potential,
$V(\vec{x})$ = overall potential field, which takes the value V_s on the satellite,
I_e = incident electron current on satellite surface,
I_i = incident ion current on satellite surface,
I_{se} = secondary electron current due to I_e,
I_{si} = secondary electron current due to I_i,
I_{be} = backscattered electrons due to I_e,
I_{ph} = photoelectron current,
I_b = active current sources such as charged particle beams or ion thrusters,
I_s = surface current to other surfaces or through the surface.

The signs of the currents are explicitly stated. Note that I_i has a negative sign since the ions are flowing to the surface and hence are antiparallel to the surface normal vector.

The key to modeling spacecraft charging is the solution of Eq. (5.15), that is, finding V_s so that the total current is zero. The basic problem is the solution of Eq. (5.15), subject to the constraints of Poisson's equation,

$$\epsilon_0 \nabla \cdot \vec{E} = -\epsilon_0 \nabla^2 V = e(n_i - n_e), \tag{5.16}$$

and the time-independent collisionless Boltzmann (or Vlasov) equation,

$$\vec{v} \cdot \nabla f_{i,e} - \frac{q_{i,e}}{m_{i,e}} \nabla V \cdot \nabla_v f_{i,e} = 0, \tag{5.17}$$

where n_e = local electron density = $\int f_e \, dv$, n_i = local ion density = $\int f_i \, dv$, and ∇, ∇_v = gradient operators with respect to position and velocity space, respectively.

The Vlasov equation arises for the same reason as the collisionless Boltzmann equation in Section 4.1. The collisions for even the dense plasmas in LEO are so low that the magnitude of the collision operator on the right-hand side of Eq. (5.17) is always small compared with the left-hand side of Eq. (5.17).

Sophisticated techniques have been developed for solving these equations. In Section 5.2.2.1, to illustrate these methods, first-order solutions of Eq. (5.15) are introduced subject to Eqs. (5.16) and (5.17). Simple, analytical approximations to the ambient currents to a probe are developed, based on the original probe theory of Langmuir and Mott–Smith. Advanced probe theories capable of modeling three-dimensional wake phenomena and the satellite photosheath also are described. Finally, a discussion of numerical simulation codes capable of explicitly modeling the detailed geometry around a space vehicle are presented.

The steady state solution to Eq. (5.13) is given by $I_{net}(V_s) = 0$. If the surface potential is close to the steady-state solution, then the timescale for it to approach steady state can be calculated in the following manner. Take V_{s0} as a steady-state solution of $I_{net} = 0$ and assume a small perturbation such that the potential is $V_s = V_{s0} + \delta V_s$, where $\delta V_s / V_{s0} \ll 1$. If the surface capacitance is C and the surface charge Q is $Q = CV_s$, then δV_s satisfies an equation generated by a Taylor-series expansion of Eq. (5.13).

$$C \frac{\partial \delta V_s}{\partial t} = \frac{dI_{net}}{dV_s}\bigg|_{V=V_{s0}} \delta V_s, \tag{5.18}$$

which has the solution

$$\delta V_s = (\delta V_s)_0 \exp(-t/\tau_r), \tag{5.19}$$

where $\tau_r = -C/(dI_{net}/dV_s)|_{V=V_{s0}}$. It is clear from Eq. (5.19) that a necessary condition for a stable solution $\delta V_s \to 0$ is that $\tau_r > 0$ or that $dI_{net}/dV_s|_{V=V_{s0}} < 0$. It is shown later that this condition can be used to distinguish between multiple solutions of $I_{net} = 0$. For a GEO spacecraft with a metal surface, $C/A \approx \epsilon_0/R$, where R is the radius of curvature of the surface, and A is its surface area. A reasonable approximation is $dI_{net}/dV_s|_{V=V_{s0}} \approx (j_a A)/(kT_e/e)$ so that $\tau_r \approx (\epsilon_0/R)(kT_e/e)/j_a$. For a metal surface in the magnetosphere, the charging time is milliseconds (Whipple, 1981). For a dielectric surface, with a large capacitance between the surface and substrate, $\tau_r \approx (\epsilon/d)(kT_e/e)/j_a$ where d is the depth of the dielectric and ϵ is the dielectric constant. This time may be as large as minutes (Whipple, 1981). Thus a spacecraft spinning at a few revolutions per minute with exposed insulating surfaces may never reach equilibrium.

A basic concept in understanding spacecraft charging is the idea of the Debye length. If it is small compared to the length of the spacecraft, then the so-called thin-sheath approximation is appropriate, whereas at the other limit, the thick-sheath approximation is more appropriate. In terms of these two approximations, analytical solutions of the charging equations are found that can be used to illustrate fundamental characteristics of charging effects.

The Debye length is introduced in Section 2.4.2. The terminology "thick" sheath and "thin" sheath derives from the assumption that the region over which the satellite affects the ambient plasma or is screened from the ambient plasma is either larger or smaller than the characteristic dimensions of the spacecraft. Usually, this reduces to determining whether λ_D is large (thick sheath) or small (thin sheath) compared to the spacecraft radius. For a 1-m-radius spacecraft with no exposed potential surfaces in LEO, the term thin sheath is appropriate. For GEO, unless the satellite is 10 m or larger, the thick sheath approximation is appropriate. If active surfaces (i.e., exposed potentials driven by the satellite systems) or a substantial photoelectron population are present, these limiting criteria will change, but even so, for most spacecraft studies they are very useful. Depending on which limit holds, specific simplifying assumptions are made that allow computations of the currents in Eq. (5.15).

5.2.2.1 The Thin-Sheath Limit

First, consider a large structure such that the characteristic scale of the sheath is significantly smaller than R_s, the radius of the surface (the thin-sheath assumption). Assume that at the surface $(X = R_s)$, the potential is V_s and the surface is nearly planar relative to the sheath dimensions. At distance $y = X - R_S$ from the surface, Poisson's equation for the attracted species becomes

$$\frac{d^2 V}{dy^2} = -\frac{qn(y)}{\epsilon_0}, \tag{5.20}$$

and the current continuity equation becomes

$$j = qn(y)v(y) = \text{const}, \tag{5.21}$$

where j is the current density, $n(y)$ is the attracted species density, q is the attracted species charge, and $v(y)$ is the attracted species velocity. If the attracted species start at zero energy far from the planar surface and accelerate without collisions toward the surface, then, by energy conservation,

$$\frac{1}{2}mv^2(y) + qV(y) = 0. \tag{5.22}$$

Combining these two equations with Poisson's equation,

$$\frac{d^2 V}{dy^2} = \frac{j}{\sqrt{-2qV/m}}\frac{1}{\epsilon_0}. \tag{5.23}$$

To solve this equation, it is assumed that V and $dV(y)/dy$ are zero at a distance $y = S$. This determines the sheath thickness and is the space-charge-limited assumption. In simple physical terms, the space charge above the surface shields the electric field originating on the surface so that at some point $(y = S)$, it goes to zero. With this assumption and after multiplying both sides of Eq. (5.23) by the factor dV/dy, the solution becomes

$$\frac{1}{2}\left(\frac{dV}{dy}\right)^2 = \frac{1}{\epsilon_0}\frac{m}{q}j\sqrt{-2qV/m}. \tag{5.24}$$

Integrating using the condition $V = V_s$ at $y = 0$ gives

$$j = \frac{4}{9}\sqrt{\frac{2q}{m}}\epsilon_0\frac{|V_s|^{3/2}}{S^2}. \tag{5.25}$$

This is the well-known Child–Langmuir law for space-charge-limited flow. It relates the current that can flow to a surface at a given surface potential to the size of the sheath, S. In the one-dimensional system considered here, sufficiently far from the surface, the current density can be equated to the random thermal current. Then, the sheath distance S can be computed self-consistently if the current density is defined by

$$j = j_0 = K^* q n_0 \sqrt{kT/m}, \qquad (5.26)$$

where $1/\sqrt{2\pi} \leq K^* \leq 1$. This follows as $K^* = 1/\sqrt{2\pi}$ for a far-field Maxwellian distribution (so that $j_0 = q n \bar{c}/4$) or $K^* = 1$ for a monoenergetic distribution with energy $E_0 = kT/2$ (so that $j_0 = q n \sqrt{2E_0/m}$). With this substitution and the definition of Debye length [see Eq. (2.54)],

$$S = \frac{2}{3} \left(\frac{\sqrt{2}}{K^*} \right)^{1/2} \lambda_D \left(\frac{|q V_s|}{kT} \right)^{3/4}. \qquad (5.27)$$

Since the sheath thickness must always be at least of the order of the Debye length, this expression for S requires that $|q V_s|/kT \gg 1$. This simple expression arises because the repelled species was neglected. This can be true only if $|q V_s|/kT \gg 1$ since the repelled species will have a density that scales as $\exp(-|q V_s|/kT) \ll 1$. In Eq. (5.27), S is sometimes called the Child–Langmuir length. The equation implies that in a highly biased situation, the electric field from the surface extends out a distance of the order of $\lambda_D (|q V_s|/kT)^{3/4} \gg \lambda_D$. For the planar assumption to hold, however, it also must be the case that $S \ll R_s$.

The sheath thickness determines the region over which charge is collected and is important in determining the maximum current that can flow to a probe for a given V. This can be understood by examining the evolution of the electric field at the sheath boundary as the incoming current and density are increased (see Figure 5.5). Initially, at very low currents, the potential profile between S and the surface is linear. As the current increases, the profile starts to flatten out, especially at $y = S$. Once the space-charge limit is reached, the electric field at S is zero and the current density is given by Eq. (5.25). If the current into the system increases, the electric field at $y = S$ reverses sign and repels the current, allowing no more particles to be injected. Therefore, the space-charge-limited current is also the maximum current that can flow to a surface for a given sheath size. This occurs for electrostatic plasma thrusters (ion engines) and is an important limiting factor in their operation. Alternatively, if the current density is fixed by the far-field current flow, the sheath is determined as specified in Eq. (5.27).

In the derivation of the Child–Langmuir law for a planar system, the repelled species was ignored in the sheath region. Although this is sometimes true in the

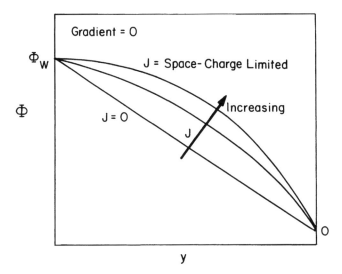

Figure 5.5. Evolution of Potential Profile as the Current in a One-Dimensional System Increases.

sheath itself, it is not valid outside the sheath, as can be seen by the following argument. Assume that species 1 is accelerated toward the wall, that species 2 is repelled from the wall, and that the plasma is quasi-neutral at the sheath edge $y = S$ (i.e., $n_1 \approx n_2$ at the sheath edge). For simplicity, consider the case where species 1 is cold ($kT_1 = 0$) but has directed velocity $v_0 \neq 0$ at the edge $y = S$. This is necessary; otherwise, $n_1(y = S)$ would have to be infinite to give a finite current. Therefore, at $y = S$, the distribution function for species 1 is

$$f_1 = n_0 \delta(v - v_0), \tag{5.28}$$

where $\delta(x)$ is the Dirac delta function. Because the plasma is taken to be collision-less, Liouville's theorem (Bittencourt, 1986) asserts that the distribution function that satisfies the Vlasov equation (5.17) must only be a function of the characteristics of the Vlasov equation. Because one of the characteristics is the particle energy, E, this implies that the distribution function for species 1 is

$$f_1(y, v) = n_0 \delta \left(\sqrt{\frac{2E}{m}} - v_0 \right) = n_0 \delta \left(\sqrt{v^2 + \frac{2q_1 V}{m_1}} - v_0 \right). \tag{5.29}$$

This distribution function reduces to Eq. (5.28) at $y = S$ and also satisfies the Vlasov equation everywhere since it is a function of one of the characteristics of the equation.

Because the repelled species 2 distribution function should be close to thermo-dynamic equilibrium, it is assumed to be

$$f_2(y, v) = n_0 \left(\frac{m_2}{2\pi k T_2} \right)^{1/2} \exp[-m_2(v^2 + 2q_2 V/m_2)/2k T_2]. \tag{5.30}$$

Dimensionless variables are defined by

$$\eta = -q_1 V/k T_2 \tag{5.31}$$

$$u = v/\sqrt{2k T_2/m_1} \tag{5.32}$$

$$\xi = y/\sqrt{\epsilon_0 k T_2/(n_0 q_1^2)}. \tag{5.33}$$

The particles of species 1 are accelerated; thus $q_1 V < 0$ so that $\eta > 0$. Note that the velocity $(2k T_2/m_1)^{1/2}$ is akin to the ion acoustic velocity (Bittencourt, 1986), and the length $[\epsilon_0 k T_2/(n_0 q_1^2)]^{1/2}$ is the Debye length in this plasma for the cold attracted species. The densities are found by integration over velocity space

$$n_1 = \frac{n_0}{1 + \eta/u_0^2} \tag{5.34}$$

$$n_2 = n_0 \exp(-\eta), \tag{5.35}$$

where u_0 is the nondimensional version of v_0. The Poisson equation becomes

$$\frac{d^2\eta}{dy^2} = \frac{1}{1 + \eta/u_0^2} - \exp(-\eta). \tag{5.36}$$

In the limit $\eta \gg 1$, this becomes a nondimensional version of Eq. (5.23). This equation can be integrated once using the space-charge-limited assumption to give

$$\frac{1}{2} \left(\frac{d\eta}{dy} \right)^2 = 2u_0^2[(1 + \eta/u_0^2)^{1/2} - 1] + [\exp(-\eta) - 1]. \tag{5.37}$$

For the requirement that $(d\eta/dy)^2 \geq 0$, a Taylor expansion around the location $\eta \approx 0$ gives

$$v_0 \geq \sqrt{k T_2/m_1}. \tag{5.38}$$

This is known as the Bohm sheath criterion. Physically, for a stable sheath to exist around a surface, the attracted species must enter the sheath with a minimum velocity. This minimum is related to the characteristics of the repelled particles. If the attracted particles do not enter with the minimum velocity, then no steady-state sheath can form and an oscillatory sheath will result. The concept that the attracted particles enter with a minimum velocity implies that these particles have been

Table 5.3. *Dependence of I/I_0 as a function of $e|V_s|/kT$ for $\lambda_D(e|V_s|/kT)^{3/4} \ll R_c$*

$e|V_s|/kT$:	0	$e|V_s|/kT \ll 1$	0.25	0.50	0.75	1.0	$e|V_s|/kT \ll (R_c/\lambda_D)^{4/3}$
I/I_0:	1	$1 + e|V_s|/kT$	1.18	1.26	1.33	1.38	1.47

accelerated by a potential drop of the order of the energy of the repelled particles. Hence, the electric field cannot become zero at a finite location but must continue out into the quasi-neutral plasma beyond the sheath at a small level. The magnitude of the field must be such that the potential drop from the field to the the edge of the sheath is of the order of the energy of the repelled particles. This is called the presheath concept. If the attracted particles are ions, then the minimum velocity is $\sqrt{kT_e/m_i}$, and the potential drop outside the sheath is kT_e/e. This minimum velocity, $\sqrt{kT_e/m_i}$, is called the Bohm velocity and is related to the ion acoustic velocity $c_s = \sqrt{2kT_e/m_i}$. The ion acoustic velocity is the velocity at which sound waves will propagate in the plasma. This wave velocity is the velocity at which waves move in order to balance the electron thermal pressure with the ion inertia.

All of the preceding derivations are for a planar system. Such a system has the advantage of being analytically tractable but the disadvantage of being unphysical in that such a system is not bounded at infinity. Al'pert et al. (1965) calculated the case of space-charge-limited collection in a plasma with $T_e = T_i = T$ and for a spherical geometry. For the case where the sheath distance $S = \lambda_D(e|V_s|/kT)^{3/4}$ satisfies $S \ll R_c$ (R_c is the radius of the sphere), and with the definition

$$I_0 = 4\pi R_c^2 \frac{n_0 \bar{c}}{4} = n_0 R_c^2 \sqrt{8\pi kT/m}, \tag{5.39}$$

the current of attracted particles is given in Table 5.3. In this equation, I_0 is the random current that would strike the sphere in a Maxwellian plasma at rest. For large potentials (but still such that the system is in the thin-sheath limit), the current collected to the sphere is enhanced by 47 percent over the random current. This is a consequence of the presheath, which in a spherical system allows the sheath field to extend into the plasma a distance farther than the sphere radius.

5.2.2.2 The Thick-Sheath Limit

The preceding theory concerning the thin sheath assumes that sheath effects dominate the current flow to the satellite. This places emphasis on Poisson's equation and the space charge around the satellite. In the opposite extreme, the sheath and space charge are ignored so that to first order, Laplace's equation holds (the right-hand side of the Poisson equation equals zero). In practical terms, this translates into the assumption that $\lambda_D \gg R_s$. For spherical symmetry, conservation of energy and angular momentum imply that for an attracted particle approaching the satellite from infinity,

$$\frac{1}{2}mv_0^2 = \frac{1}{2}mv_{R_s}^2 + qV_s,$$
$$mR_iv_0 = mR_sv(R_s),$$

(5.40)

where v_0 is the velocity in the ambient medium, R is the impact parameter, and $R = R_i$ for a grazing trajectory for a vehicle of radius R_s.

Solving for the impact parameter R_i (only particles having $R < R_i$ will reach R_s),

$$R_i^2 = R_s^2 \left(1 - \frac{2qV_s}{mv_0^2} \right),$$

(5.41)

and $(R_i - R_s) \simeq R_i$ is equivalent to the sheath thickness S defined for a thin sheath, because it also is the size of the region from which particles can be drawn.

The total current density striking the satellite surface for a monoenergetic beam is

$$j(V_s) = \frac{I}{4\pi R_s^2} = \frac{I}{4\pi R_i^2}\frac{R_i^2}{R_s^2} = j_0 \left(1 - \frac{2qV_s}{mv_0^2} \right),$$

(5.42)

where j_0 is the ambient current density outside the sheath given by $j_0 = n\bar{c}/4$.

This is the so-called "thick-sheath, orbit-limited" current relation. In sheath modeling, it represents the other extreme from the thin sheath. The density of the plasma is sufficiently low that whether or not the attracted particles are collected by the surface is determined by the orbital parameters (energy, angular momentum) of each particle in the far field. In the thin-sheath limit, the current collection is determined by the self-consistent space charge in front of the surface and does not depend on the far-field energy and momentum of the particles.

The thick-sheath results are readily extended to more complex distributions. Prokopenko and Laframboise (1980) have similarly derived in more general terms the current density to a sphere, infinite cylinder, and plane (i.e., three, two, and one dimensions) for the orbit-limited solution. Their results are, for the attracted species,

sphere

$$j = j_0(1 + Q),$$

(5.43)

cylinder

$$j = j_0[2(Q/\pi)^{1/2} + \exp(Q)\,\mathrm{erfc}(Q^{1/2})],$$

(5.44)

plane,

$$j = j_0,$$

(5.45)

and for the repelled species,

$$j = j_0 \exp(Q),\qquad(5.46)$$

where

$$Q = -qV_s/kT\qquad(5.47)$$

and $Q > 0$ for the attracted species and $Q < 0$ for the repelled species.

These effects of one-, two-, and three-dimensional geometries on charging are plotted in Figure 5.6. Several conclusions can be drawn. First, Prokopenko and Laframboise's results for a sphere are identical to Eq. (5.42) (with energy $mv_0^2/2$ replaced by kT) for a thick sheath. It also should be readily apparent that the planar solution is conceptually equivalent to a thin sheath. Thus, Eqs. (5.43) to (5.45) additionally give a qualitative picture of how the currents to the spacecraft change as the ratio of the Debye length to satellite radius is varied from small values (thin sheath or planar) to large values (thick sheath or spherical).

The case of the thick sheath is, however, of limited applicability. Al'pert et al. (1965) note that in a plasma the electric field may decrease with distance both more rapidly as well as more slowly than "$1/r^2$" but at infinity it always falls exactly as $1/r^2$. This is necessary for the electric field to be regular at infinity. It can be shown that a necessary condition for the orbit-limited current model to be valid is that the electric field fall off more slowly than $1/r^2$ up to distances of the order of $\lambda_D \gg R_s$ from the body. If the electric field falls off more quickly than this, then trapped orbits are possible and the space charge associated with those orbits can

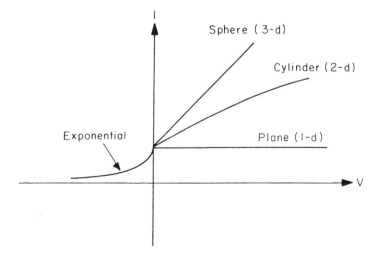

Figure 5.6. Qualitative Behavior of Current Voltage Curves for Orbit-Limited Collection.

influence the trajectories of the incoming attracted particles. Al'pert et al. (1965) show that, for a spherical system in the thick-sheath limit ($\lambda_D \gg R_s$), the attracted current is given by

$$j = j_0 \left(1 + \frac{e|V_s|}{kT}\right) \qquad e|V_s|/kT \ll (\lambda_D/R_s)^8 \tag{5.48}$$

and

$$j = 0.951 j_0 \left[\frac{e|V_s|}{kT} \left(\frac{\lambda_D}{R_s}\right)^{4/3}\right]^{6/7} \qquad e|V_s|/kT \gg (\lambda_D/R_s)^8. \tag{5.49}$$

The former case is the orbit-limited result given earlier. The latter case is space-charge limited and is obtained as

$$j = 1.47 j_0 \frac{R_{sc}^2}{R_s^2}, \tag{5.50}$$

where the space-charge-limited spherical sheath radius R_{sc} can be determined self-consistently (unlike the planar case) and is

$$\frac{R_{sc}}{\lambda_D} = 0.803 \left[\frac{e|V_s|}{kT} \frac{R_s}{\lambda_D}\right]^{3/7}. \tag{5.51}$$

Eq. (5.49) is also the equation to use if $\lambda_D \ll R_s$ but $S = \lambda_D(e|V_s|/kT)^{3/4} \gg R_s$. This applies since the potential is strong enough to pull in enough charge to affect the particle orbits rather than the particle orbits being determined only by the orbital parameters.

Several conclusions about three-dimensional charging can be drawn from the simple analysis presented. First, the sheath dimensions relative to the size of the body are very important in determining the I-V characteristics of the attracted species and, therefore, in solving the current balance equation. In the absence of space charge, the sheath can be effectively ignored for the thick sheath (i.e., the geometry of the spacecraft determines the details of the charging). For the thin sheath, the Debye shielding of the satellite potential becomes critical because the spacecraft potential drops orders of magnitude at several Debye lengths from the surface. This also severely limits the return current that the spacecraft can draw from the ambient environment. For electron emission, this significantly affects the beam behavior.

The preceding concepts find a wide range of applications in the modeling of spacecraft charging. Simple point models that employ the orbit-limited equations are commonly used to estimate the spacecraft-to-space potential for GEO spacecraft. Even though this analytical probe theory has practical applications, it does not allow for the complex geometric and space-charge effects of the satellite sheath on particle trajectories. As a consequence, it is severely limited in quantitative accuracy. To include these three-dimensional effects of the sheath on particle trajectories, it is necessary to seek numerical, self-consistent solutions of Eqs. (5.15), (5.16), and (5.17). Typically, this is not analytically possible, and an iterative procedure must be employed.

As a simplified explanation of the iterative procedure required, consider the following. Assume a spacecraft is sitting in space at a fixed potential. The potential is initially assumed to be unaffected by the plasma around the vehicle. Next, particles from the ambient environment are traced along the trajectories they follow in the spacecraft potential field. These trajectories, illustrated in terms of the "equivalent potential" $U(r, L)$ [see Bernstein and Rabinowitz (1959) and Garrett (1981)], can be divided into four types:

Type 1: The particle has sufficient kinetic energy and small enough angular momentum to reach the spacecraft surface or to go from the spacecraft surface to infinity. These particles will contribute once to a current to the spacecraft.

Type 2: The particle starts at infinity but never reaches the spacecraft because it is repelled at some minimum distance from the spacecraft. If the trajectory passes through the region of interest, however, the particle will contribute twice to the region being studied.

Type 3: The particle starts at the spacecraft but is reflected back to the surface. These particles also contribute twice to the region being studied.

Type 4: The particle is in a trapped orbit around the spacecraft.

Next, for each trajectory, depending on the type, the flux to each point on the spacecraft and the flux in the space around the spacecraft are calculated. This is used to determine the density at each point. The charge density in space and the current to the spacecraft are then used to recompute the potential field – in this way, space-charge effects are explicitly included. The process is iterated until a self-consistent solution is achieved. At that point, there is a complete description of the three-dimensional charging problem: the flux at each point on the surface and in space, the density of charged particles in space near the spacecraft, and the potentials (and electric field) on and around the spacecraft.

This type of computation, in principle, allows a self-consistent computation of spacecraft charging and current collection in three dimensions. This has been implemented in several different ways. The most straightforward numerical techniques conceptually are the so-called particle-pushing techniques. An example of a code that uses such a technique is the NASCAP (Katz et al., 1977; Rubin et al., 1980). This code is based on alternating solutions of the Poisson equation and calculations of the particle motion and charge deposition. The computation assumes that charging proceeds through a series of equilibrium states. Computations are carried out in a series of nested grids. This system permits the modeling of a volume of virtually any size desired. The code builds complex spacecraft out of simple cubes or slices through cubes. Two methods are used to calculate the charge deposition on a cell on a spacecraft surface. In the first, particles start at the spacecraft surface and are then tracked backward in time. This is based on the phase-space invariance of the Vlasov equation and is much more computationally efficient than tracking particles from the far field to the spacecraft. In the other method, a portion of the surface is locally modeled as a sphere and the differential particle fluxes to it are calculated on the basis of the orbit-limited theory outlined above. NASCAP includes secondary electron currents as well as photoelectron currents and has detailed material properties for many materials used in space. Since NASCAP implicitly uses orbit-limited theory for particle tracking, it is primarily applicable to GEO spacecraft where space-charge effects usually can be ignored. In Section 5.2.3, sample charging calculations using NASCAP or the simple models described earlier are presented for GEO spacecraft.

5.2.3 Spacecraft Potentials at GEO

For any surface on a spacecraft in the space environment, the current-balance equation states that the net current to the surface must be zero at equilibrium. The surface of a spacecraft is composed of a mosaic of elements made from many different materials (Grard, 1983). Individual elements can be complex in both configuration and composition. Solar cells, for example, consist not only of the semiconductor material but have insulating coverglasses for radiation protection. Thermal blankets may have multiple layers of aluminized Kapton or Mylar. Surface radiators may be made of quartz or other insulators. The characteristic thickness of insulated surfaces is between 10 and 50 μm with material resistances of 10^{17}–10^{18} Ω. Painted surfaces have resistances of the order of 10^{12} Ω. Modern charge-control design practices for GEO spacecraft require that all conductive surfaces be grounded to the spacecraft bus, which is therefore the reference ground for all of the electronic and power modules. Although a complete calculation of the potential on a spacecraft requires careful consideration of the potential on each surface and the current flow between the surfaces, valuable insights can be gleaned from analytical solutions to the current-balance equation for a single point. This assumes, however, that all of the currents can actually reach or escape from the given point on the spacecraft. At GEO, this may not be true because the ambient Debye length is much larger then typical spacecraft sizes. Therefore, one highly biased part of the spacecraft may electrically affect current collection to other parts of the spacecraft. When this is the case, currents everywhere on the spacecraft will be affected and a complete numerical calculation becomes necessary.

A simple case that allows physical insight into the charging process is that of a surface in darkness with no secondary emission and no backscatter. For an ambient plasma consisting of two single Maxwellian populations, one electrons and the other protons, the net, orbit-limited current density (flux) to a sphere, from Eqs. (5.43) and (5.46), is

$$j_{\text{net}}(V_s) = j_{0_e} \exp(-e|V_s|/kT_e) - j_{0_i}(1 + e|V_s|/kT_i) \quad V_s < 0 \tag{5.52}$$

and

$$j_{\text{net}}(V_s) = j_{0_e}(1 + eV_s/kT_e) - j_{0_i} \exp(-eV_s/kT_i) \quad V_s > 0. \tag{5.53}$$

For $V_s = 0$,

$$j_{\text{net}} = j_{0_e} - j_{0_i} \propto \left(\sqrt{T_e/m_e} - \sqrt{T_i/m_i} \right). \tag{5.54}$$

Because $m_e \ll m_i$ and for $T_i \gg T_e$, it can be seen that $j_{\text{net}} \approx j_{0_e} > 0$. Furthermore, from Eq. (5.53), $dj_{\text{net}}/dV_s > 0$. No root for $j_{\text{net}} = 0$ can be found when $V_s > 0$. Hence, the solution of $j_{\text{net}} = 0$, from Eq. (5.52), is (for $V_s < 0$)

$$V_s = -\frac{kT_e}{e} \ln \left[\sqrt{\frac{T_i}{T_e} \frac{m_i}{m_e}} (1 - eV_s/kT_i) \right]. \tag{5.55}$$

For protons, $\sqrt{m_i/m_e} \approx 43$, and for $T_e \approx T_i$ the solution for V_s is $V_s \approx -2.5kT_e/e$. The surface potential for zero net current collection is thus of the order of and scales as the electron temperature. This can be interpreted physically as follows. For zero surface potential relative to space, the higher mobility of the electrons relative to the protons means that net current to the surface is dominated by the electron current. To reduce this electron current to the level of the proton current (i.e., for current balance), the surface potential must be negative to repel the electrons and increase the proton current. Since the electron current is much larger than the proton current for zero surface potential, negative charge will collect on the surface. The surface potential, therefore, will become increasingly negative, repelling more and more of the electron current. The (negative) potential, which scales as the electron temperature, will become more negative until a current balance is achieved that gives net zero current for the incoming electrons and incoming protons.

The simple model just presented is capable of generalization to cover much more complex cases where secondary emission and multiple Maxwellians are allowed. If a surface in a multiple Maxwellian plasma (with no photoemission, secondary emission, or backscattering) is considered, then for $V_s < 0$ the net current density is

$$j_{\text{net}}(V_s) = \sum_{k=1}^{k=2} \left[j_{0_{e_k}} \exp(-e|V_s|/kT_{e_k}) - j_{0_{i_k}}(1 + e|V_s|/kT_{i_k}) \right] \quad V_s < 0. \tag{5.56}$$

For $n_{e_1} \approx n_{e_2}$ and with the ordering $T_{e_2} > T_{e_1}$, the electron current density associated with the temperature T_{e_2} is larger than the electron current density associated with T_{e_1}. To achieve current balance, the surface potential in this case must repel the high-energy electron component, giving a surface potential $V_s \approx -kT_{e_2}/e$ [see Eq. (5.55)]. This suggests that a spacecraft surface in darkness at GEO may achieve potentials of the order of $-10,000$ to $-30,000$ V (i.e., proportional to the high-energy electron temperature).

If photoelectrons are next included, the net current is

$$j_{\text{net}}(V_s) = \sum_{k=1}^{k=2} \left[j_{0_{e_k}} \exp(-e|V_s|/kT_{e_k}) \right.$$

$$\left. - j_{0_{i_k}} (1 + e|V_s|/kT_{i_k}) \right] - j_{\text{ph}_0} \quad V_s < 0. \tag{5.57}$$

and

$$j_{\text{net}}(V_s) = \sum_{k=1}^{k=2} \left[j_{0_{e_k}} (1 + eV_s/kT_{e_k}) - j_{0_{i_k}} \exp(-eV_s/kT_{i_k}) \right]$$

$$- j_{\text{ph}_0} \exp(-eV_s/kT_{\text{ph}})(1 + eV_s/kT_{\text{ph}}) \quad V_s > 0, \tag{5.58}$$

where, from Section 5.2.1.2, the photocurrent for zero potential is j_{ph_0} and kT_{ph} is the mean energy with which the photoelectrons leave the surface. From Section 5.2.1.2, as $j_{\text{ph}_0} \gg j_{0_{e_{1,2}}}$, $V_s = 0$ implies that the net current is negative. For $V_s \to \infty$, the net current in Eq. (5.58) is positive. Therefore, it must be the case that a root of $j_{\text{net}} = 0$ exists for $V_s > 0$. Thus, from Eq. (5.58),

$$V_s \approx \frac{kT_{\text{ph}}}{e} \ln \left[j_{\text{ph}_0} (1 + eV_s/kT_{\text{ph}})/j_{0_{e_2}} \right], \tag{5.59}$$

because it is likely that $kT_{\text{ph}} \ll kT_{e_2}$. From this it can be seen that $V_s \sim kT_{\text{ph}}/e$. The current balance is between the outgoing photoelectrons and the incoming hot electrons. The surface potential must float positive relative to the space potential to increase the incoming electron current because the emitted photoelectron current is so much larger than the ambient electron current. It must rise to the order of the mean energy with which the photoelectrons are ejected. Therefore, a surface in sunlight may have a potential of the order of $+2$–5 V. However, this result must be treated with caution because it assumes that there is no external barrier field inhibiting the outflow of photoelectrons. This is not usually true, as will be seen in the Section 5.2.3.1.

The next case to be considered is for a surface with secondary electron emission and in darkness. (*Note:* Backscatter will not be considered. It can be included as a special case of multiple Maxwellian plasmas because the backscattered current primarily acts to reduce the incoming electron current by some fraction). If Eq. (5.4) is integrated over the incoming Maxwellian distribution functions and normal incidence for the electrons is assumed, the net current density (Prokopenko and Laframboise, 1980; Hastings, 1986) is

$$
j_{\text{net}}(V_s) = \sum_{k=1}^{k=2} \left\{ j_{0_{e_k}} \exp\left(-e|V_s|/kT_{e_k}\right) \right.
$$

$$
\left. \times \left[1 - \bar{\delta}\left(\frac{kT_{e_k}}{E_{\max}}\right) \right] - j_{0_{i_k}}\left(1 + e|V_s|/kT_{i_k}\right) \right\} \quad V_s < 0 \quad (5.60)
$$

and

$$
j_{\text{net}}(V_s) = \sum_{k=1}^{k=2} \left[j_{0_{e_k}}\left(1 + eV_s/kT_{e_k}\right) - j_{0_{i_k}} \exp\left(-eV_s/kT_{i_k}\right) \right.
$$

$$
\left. - j_{0_{e_k}} \bar{\delta}\left(\frac{kT_{e_k}}{E_{\max}}\right) \exp(-eV_s/kT_{\text{sec}})(1 + eV_s/kT_{\text{sec}}) \right] \quad V_s > 0.
$$

$$
(5.61)
$$

In these equations, kT_{sec} is the energy with which the secondary electrons leave the surface. The function $\bar{\delta}(x)$ [where $x = (kT_{e_k}/E_{\max})$] is the integral of the secondary yield for normal incidence over a Maxwellian distribution function. It is approximated by

$$
\bar{\delta}(x) = 7.4\delta_{e_{\max}} \left\{ x^3 + \frac{9}{2}x^2 + 2x \right.
$$

$$
\left. - x^{3/2} \exp(x)\left(x^2 + 5x + \frac{15}{4}\right)\sqrt{\pi}[1 - \text{erf}(\sqrt{x})] \right\}. \quad (5.62)
$$

It has the properties $\bar{\delta}(x) \sim x$ for $x \ll 1$, $\bar{\delta}(x) \sim 1/x$ for $x \gg 1$, and assumes a maximum value of $0.88\delta_{e_{\max}}$ for $x \simeq 0.55$. At zero potential, Eq. (5.61) gives

$$
j_{\text{net}}(0) = \sum_{k=1}^{k=2} \left[j_{0_{e_k}} - j_{0_{i_k}} - j_{0_{e_k}} \bar{\delta}\left(\frac{kT_{e_k}}{E_{\max}}\right) \right]. \quad (5.63)
$$

For $j_{\text{net}}(0) > 0$, these arguments show that the surface potential has to be negative. For $j_{\text{net}}(0) < 0$, the same arguments as for photoelectron emission imply that a positive surface potential is possible. By analogy with the photoelectrons, the surface potential in this case will scale as $V_s \sim kT_{\text{sec}}/e$. The current balance will be between incoming hot electrons and outgoing cold secondary electrons. It is

Table 5.4. *Critical temperature*

Material	T_* (eV)
Kapton	500
Teflon	1,400
Cu–Be	1,300
Cu–Be (activated)	3,700
Silver	1,200
Gold	2,900
Magnesium oxide	2,500
Silicon dioxide	1,700

clear from the discussion that $j_{net}(0) = 0$ represents a boundary between positive and negative charging of the spacecraft surfaces. For a single Maxwellian plasma and ignoring the small ion contribution, Eq. (5.61) becomes

$$0 = j_{0_e} \left[1 - \bar{\delta} \left(\frac{kT_e}{E_{max}} \right) \right]. \tag{5.64}$$

This equation suggests the definition of a material-dependent critical or threshold temperature, T_*, such that

$$\bar{\delta} \left(\frac{kT_*}{E_{max}} \right) = 1. \tag{5.65}$$

In a single Maxwellian plasma, if the ambient electron temperature exceeds this critical temperature, then negative surface charging is possible. If the ambient electron temperature is less than the critical temperature, then the spacecraft surface may charge positively. Note that if Eq. (5.65) is satisfied at all, then it will be satisfied for two values of T_* (see Figure 5.2). However, for all real materials, the first unity crossing occurs at very low energy, typically 40–80 eV. This is lower than the typical value of the electron energy at GEO. Therefore, the critical temperature is taken as the second unity crossing with the property that $\bar{\delta}(\frac{kT_*^-}{E_{max}}) > 1$ and $\bar{\delta}(\frac{kT_*^+}{E_{max}}) < 1$. For a slow change in the ambient electron temperature through T_*, there will be a large and abrupt change in the surface potential. This is illustrated by data from ATS-5 and *ATS-6* where the spacecraft potential to space is plotted relative to the mean electron temperature (Garrett, 1981). In Figure 5.7, the spacecraft potential relative to space is very small until a critical temperature of about 1,500 eV is reached. It then becomes negative and scales as the electron temperature as predicted in the discussion above. The critical temperature is estimated for a number of materials in Table 5.4 (Lai, 1991).

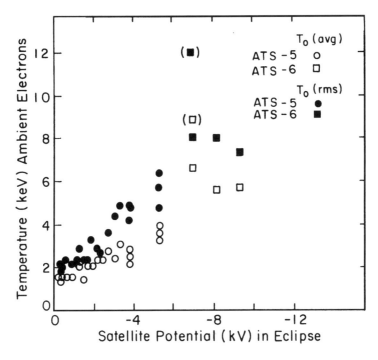

Figure 5.7. ATS-5 Spacecraft Charging Data. From Garrett, *Reviews of Geophysics*, vol. 19, pp. 577–616, 1981, Copyright by the American Geophysical Union.

Under some circumstances, a spacecraft surface may have a choice of whether it is positive or negative. This occurs when the current-balance relation allows not just one solution but three solutions (Prokopenko and Laframboise, 1980; Hastings, 1986; Lai, 1991b). Lai (1991b) shows that a necessary and sufficient condition for the existence of a triple solution is

$$j_{net}(0) < 0 \qquad\qquad (5.66)$$

$$\max[j_{net}(V_s)] > 0 \quad V_s < 0. \qquad (5.67)$$

This is illustrated in Figure 5.8 from Lai (1991b). (*Note:* As the flux is plotted rather than current density, the results must be multiplied by -1 to get the results discussed above.) Three roots are shown. The middle root is always unstable since $dj_{net}/dV_s|_{V=V_{s0}} < 0$. The discussion associated with Eq. (5.19) shows that this leads to exponential divergence away from this root. The existence of two stable roots under certain conditions will lead to ambiguity as to which solution a surface will choose. This situation of bistable equilibria is analogous to the classical van der Waals theory of phase transitions (Nicolis and Prigogine, 1977). Each stable equilibrium can be interpreted as a stable phase. The jump from one to the other can be interpreted as a phase transition. If each phase is free from fluctuations, then the

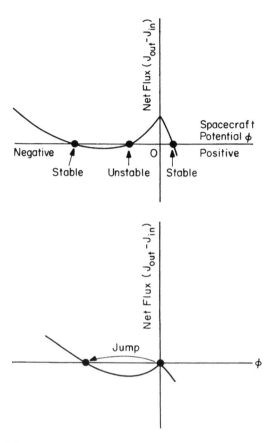

Figure 5.8. Flux Voltage Behavior in a Triple Root Situation. From Lai, *Journal of Geophysical Research*, vol. 96, pp. 19, 269–81, 1991a, Copyright by the American Geophysical Union.

initial conditions determine which root will be found. For example, if a surface initially is positively biased, as conditions change to where two bistable solutions exist, the surface will stay positively biased as long as a positive root exists. However, in the presence of sufficient fluctuations in the ambient parameters, a free-energy function akin to the Gibbs free energy can be defined that always uniquely determines the most probable solution (phase) for any environmental conditions (Hastings, 1986).

In Figures 5.9, 5.10, and 5.11, Lai (1991b) demonstrates this multiroot phenomenon using estimates of the two Maxwellian electron densities and temperatures for day 114, 1979, from the *SCATHA* satellite. The hot-electron density and temperature are relatively constant whereas the cooler density and temperature change considerably with time. *SCATHA* had a boom with an end made of an alloy of copper and beryllium (Cu–Be). The alloy is assumed to be a secondary electron emitter with $\delta_{e_{max}} = 5$ and $E_{max} = 400\,\text{eV}$. Assuming that the hot density is $0.9\,\text{cm}^{-3}$ and the solution for the hot temperature is $24.8\,\text{keV}$, the solutions for the regions of

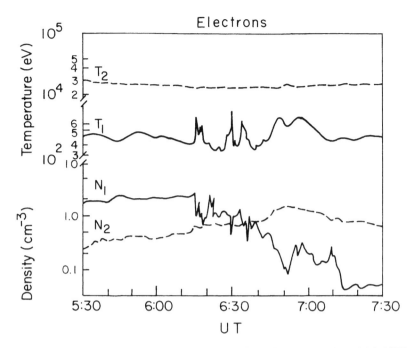

Figure 5.9. Double Maxwellian Parameters for Ambient Electrons on Day 114, 1979, from *SCATHA* Data. From Lai, *Journal of Geophysical Research*, vol. 96, pp. 19, 269–81, 1991a, Copyright by the American Geophysical Union.

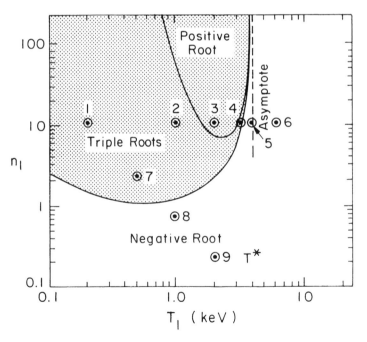

Figure 5.10. Parametric Domains for *SCATHA* Data and for Cu–Be. From Lai, *Journal of Geophysical Research*, vol. 96, pp. 19, 269–81, 1991a, Copyright by the American Geophysical Union.

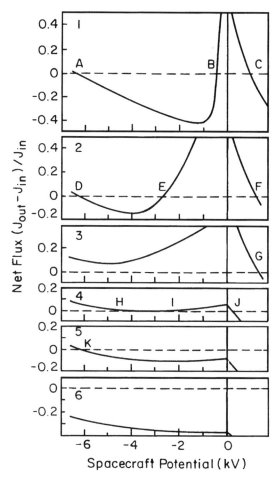

Figure 5.11. Flux Potential Behavior at Various State Points 1–6 in Figure 5.10. From Lai, *Journal of Geophysical Research*, vol. 96, pp. 19, 269–81, 1991a, Copyright by the American Geophysical Union.

triple roots can be obtained in terms of the cooler density and temperature. These are shown in Figure 5.10 where regions of negative potential, positive potential, and triple roots are plotted. The critical temperature corresponds to the region where the triple roots just appear. In Figure 5.11, the flux potential behavior is plotted for the state points 1–6 identified on Figure 5.10. The theory predicts that the potential on the Cu–Be surface starts with three possibilities but drops to one highly negative solution.

5.2.3.1 Barrier Potentials

The simple point calculations of the surface potential discussed in the preceding section offer useful physical insight but cannot reliably give surface potentials on

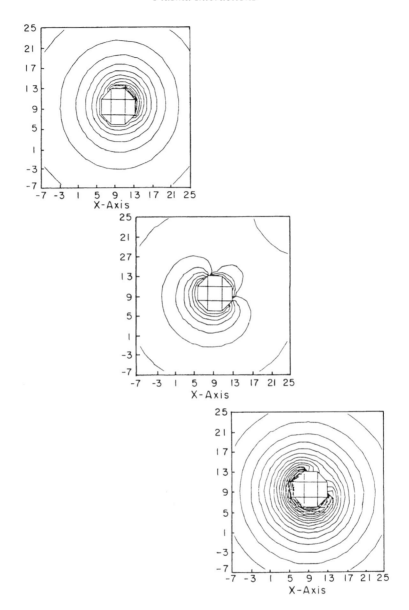

Figure 5.12. Formation of a Potential Barrier. Potentials about a Sunlit Kapton Sphere at 0.024 s, 0.5 s, and at Equilibrium. The Sun is in the Upper Right-Hand Corner.

complex surfaces. This is illustrated by a NASCAP simulation of the charging of a sunlit Kapton sphere at GEO. (Kapton is a nonconducting surface material often used for thermal control on spacecraft.) In Figure 5.12, the potential contours are plotted around the Kapton sphere for several different times. Initially, at zero seconds, the sunlit side starts at 2 V positive with current balance between incoming electrons and outgoing photoelectrons. The negative side charges negatively. After a half a second, the dark side is at -8 V and the sunlit side is at 3 V. However, a potential barrier of $+1$ V has formed over the sunlit side because of the presence of the nearby positively biased surface and the bipolar structure of the electric field [see Besse and Rubin (1980)]. As time progresses, the potential barrier grows, but at some point the low-energy photoelectrons can no longer escape. Once this happens, the entire sphere charges negatively and reaches equilibrium in about 1,000 s with a satellite ground potential of -18 kV. This complex charging behavior illustrates the importance of three-dimensional effects in modeling spacecraft charging – an issue pursued in Section 5.2.4.

5.2.4 Potentials, Anomalies, and Arcing on GEO Spacecraft

In the preceding section, the modeling of isolated surfaces on spacecraft at GEO is developed. However, because of three-dimensional effects, real spacecraft require three-dimensional charging codes such as NASCAP. In this section, model predictions are presented for the *SCATHA* spacecraft. The *SCATHA* spacecraft, launched in January 1979, was a spin-stabilized satellite that was placed in a near-geosynchronous, near-equatorial Earth orbit. It carried a variety of instruments to measure surface potentials, discharge transients, and the environment. The results from *SCATHA* have been documented extensively in the archival literature (Adamo and Matarrese, 1983; Koons, 1983; Gussenhoven and Mullen, 1986; Mullen et al., 1986; Craven, 1987; Koons et al., 1988; Koons and Gorney, 1991). Figure 5.13 is a sketch of the spacecraft. A NASCAP model of *SCATHA* is shown in Figure 5.14.

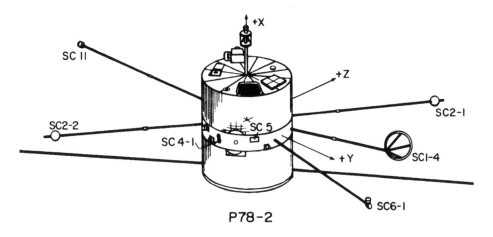

Figure 5.13. The *SCATHA* Spacecraft.

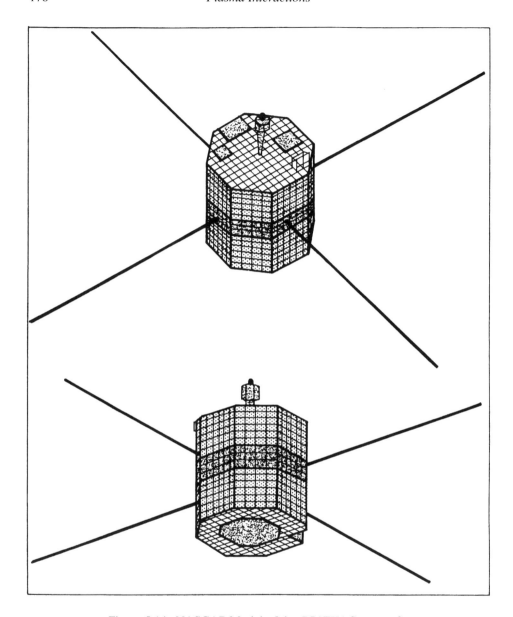

Figure 5.14. NASCAP Model of the *SCATHA* Spacecraft.

SCATHA had six distinct underlying conductors, namely, the spacecraft ground, the reference band around the belly of the spacecraft, and four experimental mountings called SC2-1, SC2-2, SC6-1, and SC6-2. In addition, most of the surface was covered with solar cells that expose coated fused silica to the space environment. A NASCAP analysis of the effects of the environment described in Table 3.13

demonstrates the complex charging behavior of each of these surfaces. Calculations show that, after 20 minutes of charging in the severe substorm environment, the solar cell coverglasses reach a voltage of $-15,600$ V. The spacecraft ground, however, is at $-15,200$ V. NASCAP also illustrates two types of differential charging that can occur on the spacecraft. The first is a differential potential between the surface of an insulator and the underlying conductor. The second is that due to the potential gradient along and between surfaces. The NASCAP run predicts that the highest electric fields internal to an insulator due to the first type of differential charging are within the Kapton surfaces just below the bellyband. The Kapton surface charges to $-16,400$ V, but the underlying conductor only reaches $-15,200$ V. The internal field is predicted to be $\sim 10^7$ V/m, which could cause dielectric breakdown of the Kapton. The second type of differential charging occurs on the booms where the material (which is platinum-banded Kapton) keeps the surface potential near the space potential. Right at the end of the booms, however, the surface potential is predicted to jump to $-22,900$ V creating a very large potential gradient along the booms.

The Russian space program has developed the ECO-M and KULON charging codes, which are similar to NASCAP. Because many Russian satellites are on 12-hour Molniya orbits, these codes also contain the capability to model radiation-induced discharges. The combination of differential charging, arcing, and radiation-induced charging has been found to be design limiting for many Russian satellites.

The accumulation of charge on or in the surface of a spacecraft can lead to spacecraft anomalies, presumably through arc discharges. Spacecraft anomalies include clock resets, power-on resets, uncommanded mode changes in instruments, and so forth. Historically, the results of one of the first studies of these anomalies is shown in Figure 5.15, which plots anomalies relative to local time. (*Note:* Radial distance has no significance in this figure.) The population of anomalies in this plot can be divided into two groups. One group is concentrated between midnight and dawn local time. This region overlaps the energetic, high-density plasma that is injected from the Earth's magnetospheric tail and convected eastward around the Earth. It is also the region that is expected to have the highest spacecraft-to-space potentials and, presumably, differential potentials (though the latter is very dependent on geometry). Enhanced discharge rates are thus also anticipated in this region. The second group of upsets is uniformly distributed around the Earth and is currently believed to be due to arcs resulting from deep dielectric charging of exposed cables by the energetic electron population at GEO (Vampola, 1987). These energetic electrons irradiate cables and lead to large internal fields that eventually relax by releasing charge through arc discharges. This is discussed in Section 6.4.

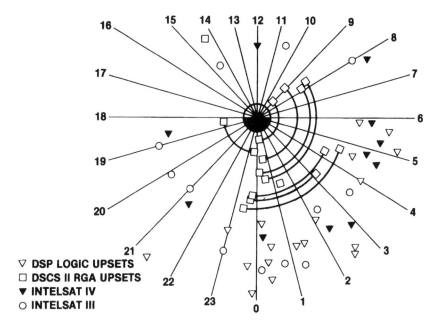

Figure 5.15. A Local Time Plot of the Occurrence of Anomalies on Various Operational Spaceraft [Rosen (1976). Copyright © 1976 AIAA. Reprinted with Permission.]

The realization that charge buildup on and in GEO satellites can lead to many problems has led to the development of design guidelines for such satellites (Purvis et al., 1984). These include:

- All conducting elements, surface and interior, should be tied to a common electrical ground, either directly or through a charge bleedoff resistor.
- For differential charging control, all spacecraft exterior surfaces should be at least partially conductive.
- The primary spacecraft structure, electronic component enclosures, and electrical cable shields should provide a physically and electrically continuous shielded surface around all electronics and wiring.
- Electrical filtering should be used to protect circuits from discharge-induced upsets.

Spacecraft designed using these guidelines have proven to be relatively immune to upsets associated with arc discharges. The *SCATHA* satellite is an excellent example of this – despite numerous, potentially damaging arcs, the spacecraft suffered few deleterious effects.

5.3 Plasma Flow Around a LEO Spacecraft

Table 5.5 presents the plasma parameters for the *space station* orbit based on the environment of Table 3.8. In contrast to GEO, the LEO environment plasma

Table 5.5. *Plasma parameters for the* space station *orbit*

Parameter	Maximum		Minimum		Average	
	e	*i*	*e*	*i*	*e*	*i*
$T_{e,i}$ (eV)	0.23	0.13	0.07	0.05	0.12	0.09
$v_{\mathrm{th}_{e,i}}$ (m/s)	2.8×10^5	1.8×10^3	1.5×10^5	1.1×10^3	2.1×10^5	1.5×10^3
$\omega_{p_{e,i}}$ (Hz)	1.5×10^7	1.2×10^5	1.4×10^6	1.1×10^4	5.2×10^6	4.3×10^4
$\Omega_{e,i}$ (Hz)	1.2×10^6	8.5×10^1	5.4×10^5	3.6×10^1	7.9×10^5	5.4×10^1
$\rho_{e,i}$ (m)	8.4×10^{-2}	7.7	2×10^{-2}	2.1	4.2×10^{-2}	4.3
λ_d (m)	2.3×10^{-2}		0.1×10^{-2}		0.4×10^{-2}	
ν_{ee} (Hz)	7.8×10^3		1.1×10^1		4.0×10^2	
ν_{ii} (Hz)	8.9×10^1		0.2		5.2	

interactions are intimately tied to the characteristics of the ambient plasma flow. Employing the plasma nondimensional parameters developed in Section 2.4, the plasma interaction with the *space station* is seen to be mesosonic; magnetized for both electrons and ions; collisional for the electrons, collisionless for the ions; and quasi-neutral on the scale of the body.

The significance of each of these factors is investigated in the following with emphasis on the anisotropy that the plasma flow introduces into the modeling process. As a consequence, the modeling of the LEO plasma environment can be much more complex than that of the GEO.

5.3.1 The Plasma Wake Structure of a LEO Spacecraft

Section 4.1 describes the neutral gas flow about a spacecraft in LEO. The plasma flow around a LEO spacecraft has the same general characteristics as the neutral flow. There is a region of compression in the ram and a wake region behind the body. The wake region, however, has a very different structure because of the electrical and magnetic forces that can influence the motion of the particles. (These electrical and magnetic forces can arise from the ambient medium or be generated by the spacecraft.) A case of particular interest is the structure and current collection in the wake of a highly biased spacecraft such as might occur for a high-power *space station*. The salient features of the plasma wake (Al'pert, 1983) are the following:

- Immediately behind the body, the electron and ion densities will be much larger than if the particles were treated as neutral particles. The electron density will be much larger than the ion density.
- Behind the body, focusing of charged particles will occur. The region of maximum rarefaction lies on a conical surface with an opening angle of $\sin^{-1}(c_s / V_0)$. This is akin to a Mach cone behind a supersonic aircraft.

- Under some conditions, the focusing of the charged particles behind the body may exceed the ambient density.
- The focusing effect depends strongly on the potential of the back surface of the spacecraft as well as the temperature ratio T_e/T_i.
- At great distances from the body, but slightly off axis, two enhanced regions may appear.
- Under the influence of an ambient magnetic field, the structure of the far-field wake becomes smoothed at a distance of the order of V_0/Ω_{p_i}.

The general analytical solution of the plasma wake behind a complex spacecraft is impossible. However, considerable insight can be gained from the analytical solution of the plasma flow around a flat plate (Wang, 1991). For the conditions in LEO, consider a two-dimensional plate in a cold, collisionless plasma flow. If the potential on the plate, $\Phi_w(\vec{r}, t)$, is less than or equal to 0, then the electrons in the plasma will always be repelled and will be Boltzmann distributed (see Eq. (2.32) and associated discussion). The potential in the near field of the plate will be

$$\nabla^2 \Phi = -(e/\epsilon_0)[n_i - n_0 \exp(e\Phi/kT_e)], \qquad (5.68)$$

and the ion density for the cold ions will be determined from continuity

$$(Dn_i/Dt) = 0 \qquad (5.69)$$

and momentum balance

$$(D\vec{v}_i/Dt) = -(e/m_i)\nabla \Phi, \qquad (5.70)$$

where D/Dt is the total derivative.

At zero or near-zero angle of attack, the plasma flowfield can be divided into three regions (see Figure 5.16). The quasi-neutral region is labeled region III. The sheath region is labeled either region I (the leading edge) or region II (the fully developed sheath layer). A final region is the wake region. In the quasi-neutral region, a combination of the Poisson equation, quasi-neutrality, ion continuity, and momentum balance gives

$$\nabla \cdot (n_i m_i \vec{v}_i) = 0 \qquad (5.71)$$

$$n_i m_i (\vec{v}_i \cdot \nabla)\vec{v}_i + \nabla(n_i kT_e) = 0. \qquad (5.72)$$

These two equations are the same as the continuity and momentum equations for a compressible gas flow (Vincenti and Kruger, 1965) with pressure $p = n_i kT_e$. Therefore, the plasma flow in region III can be analyzed as a supersonic gas flow problem.

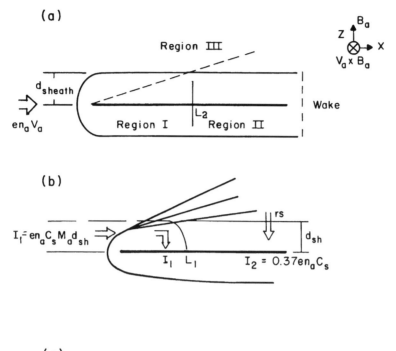

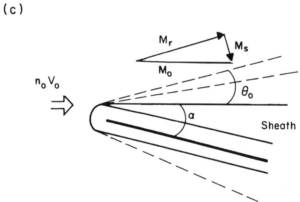

Figure 5.16. Plasma Flowfield Around a Plate at Zero or Near Zero ($\alpha < \theta_0$) Angle of Attack. [From Wang (1991).]

In region II, the ions must enter the sheath and satisfy the condition

$$M_s = (v_z^s/c_s) \geq 1. \tag{5.73}$$

In this equation, v_z^s is the normal velocity at the sheath edge. This is the Bohm sheath condition once again. The flux of ions into the sheath from region III is $\Gamma^s = n_i^s v_z^s = n_i^s c_s$. This is a sink of ions for region III. With this boundary condition, the solution for region III is identical to a Prandtl–Meyer expansion of a supersonic gas flow over a convex corner (Vincenti and Kruger, 1965). The outer envelope of the expansion fan is at $\theta_0 = \sin^{-1}(1/M_0)$, where the Mach number $M_0 = V_0/c_s$. The rarefaction is described by the relation

$$dn_i = \frac{M^2}{\sqrt{M^2 - 1}} n_i d\theta. \tag{5.74}$$

This integrates to

$$\ln \frac{n_0}{n_i(\theta)} = -\frac{e\Phi}{kT_e} = \theta \sqrt{M_0^2 - 1} + \frac{1}{2}\theta^2. \tag{5.75}$$

The turning of the ion trajectories is given by

$$d\theta = \sqrt{M^2 - 1}(dM/M). \tag{5.76}$$

The expansion fan from the edge of the plate is the presheath. Disturbances in the presheath are ion sound waves. As an ion travels through the presheath, its trajectory is turned toward the plate in such a way that its normal velocity is always sonic. From the Bohm sheath criterion and with the use of Eq. (5.76), the sheath edge potential is $\Phi_s = -kT_e/e$, and the ion density at the sheath edge is $n_0 \exp(e\Phi_s/kT_e) = 0.37n_0$. Therefore, the ion current to the plate in region II is independent of the plate potential and is $j_i = e\Gamma^s = 0.37n_0c_s$. The sheath thickness can be shown to scale as

$$\frac{d_{sh}}{\lambda_D} = 1.3|\Phi_w|^{3/4}. \tag{5.77}$$

For nonzero angles of attack (where the angle of attack α is measured clockwise from the flow direction), the analysis for a semiinfinite plate will be undertaken in polar coordinates centered on one of the corners (Lam and Greenblatt, 1966). Since the flow is irrotational, it can be solved by means of a velocity potential. Lam and Greenblatt (1966) show that for a plate in the quasi-neutral approximation (i.e., the sheath region is neglected), the ion flux to the front surface of the plate is

$$\Gamma_f(\alpha) = n_0 c_s F_f(\alpha) \qquad\qquad \alpha < \theta_0 \tag{5.78}$$
$$\Gamma_f(\alpha) = n_0 c_s M_0 \sin\alpha \qquad\qquad \alpha > \theta_0, \tag{5.79}$$

whereas the ion flux to the wake-side surface is

$$\Gamma_w(\alpha) = n_0 c_s F_w(\alpha), \tag{5.80}$$

and the functions F_f and F_w are given by

$$F_f(\alpha) = \exp\left[-\sqrt{M_0^2 - 1}(\theta_0 - \alpha) - \frac{1}{2}(\theta_0 - \alpha)^2\right] \qquad (5.81)$$

and

$$F_w(\alpha) = \exp\left[-\sqrt{M_0^2 - 1}(\theta_0 + \alpha) - \frac{1}{2}(\theta_0 + \alpha)^2\right]. \qquad (5.82)$$

At zero angle of attack and for $M_0 = 8$, $\Gamma_w = \Gamma_f = 0.37n_0c_s$, as before. As the angle of attack increases, $\Gamma_f/(n_0c_s)$ increases to one at $\alpha = \theta_0 = 7.1°$. Meanwhile, $\Gamma_w/(n_0c_s)$ decreases to 0.13 for $\alpha = \theta_0 = 7.1°$. The rarefaction angle is defined as $\delta\theta = |\theta_0 - \theta|$. This satisfies Eq. (5.75). For $M_0 = 8$, it can be shown that $\delta\theta(n_i/n_0 = 0.1) = 16.3°$ and $\delta\theta(n_i/n_0 = 10^{-4}) = 62°$. Therefore, for a mesothermal plasma flow in the quasi-neutral approximation, the near wake region behind the plate is a near vacuum.

For the case where the plate is biased and the sheath cannot be ignored, the wake structure must be found numerically (Wang, 1991). Figure 5.17 is the output of a detailed two-dimensional particle-in-cell calculation of a plasma flow over a plate at zero angle of attack. The plate length is 400 Debye lengths, the flow Mach number is 8, and the surface potential is $\Phi_w = -20kT_e/e$. The potential structure shows an expansion fan centered on the leading edge of the plate and composed of ion acoustic waves. The macroscopic ion flux shows the wake region and smoothly transitions back to the uniform flow in a few plate lengths. The ion trajectories demonstrate that ions are pulled into the wake region. In Figure 5.18, the wake structure is shown for a plate with a surface potential of $\Phi_w = -80kT_e/e$ and two angles of attack. For the larger angle of attack, there are two ion streams that are pulled around into the wake region. These are due to ions that just pass over the edge of the plate and receive an impulsive kick into the wake (Wang, 1991). The impact of these ion streams on the back of the plate can lead to sputtering from the plate. The local current density associated with these ion streams can exceed the ambient ion current density.

The wake structure around a biased plate in a mesothermal flow can be defined qualitatively in terms of two parameters. These are the nondimensional potential $e\Phi_w/kT_e$ and the nondimensional sheath ratio d_{sh}/L_t, where L_t is the plate length and $d_{sh}/L_t = 1.3|e\Phi_w/kT_e|^{3/4}/(\sqrt{M_0}\sin\alpha L_t/\lambda_D)$. The first is a measure of the effective energy that the plate can impart to a particle as it passes the plate. The second is a measure of how far the plate can reach transverse to the edge in modifying the ion trajectories. In Figure 5.19, the transition in the wake structure is shown from the quasi-neutral limit, where $e\Phi_w/kT_e \ll 1$, to the high potential limit, where $e\Phi_w/kT_e \gg 1$. This can be divided into a space-charge-limited regime for $d_{sh}/L_t \ll 1$ and an orbit-limited regime for $d_{sh}/L_t \gg 1$.

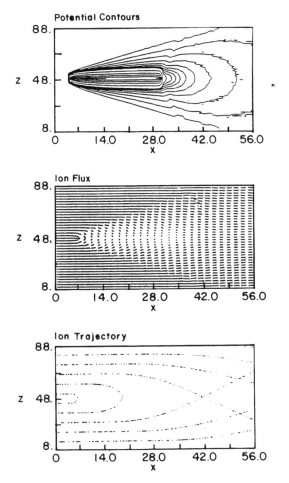

Figure 5.17. Near-Wake Region of a Flat Biased Plate in a Plasma Flow. [From Wang (1991).]

5.3.2 Current Collection in Flowing Magnetoplasmas

In previous sections, the theory of current collection in a collisionless, stationary unmagnetized plasma was introduced. Although this is relevant theory for current collection at GEO, for LEO or polar orbits, the current collection to the spacecraft takes place in a regime where the electrons are magnetized and the ions are supersonic. The relevant theory is then that associated with current collection in a flowing magnetoplasma (Laframboise and Sonmor, 1993).

The collection of ions to an attractive surface in the ram for small-to-moderate potentials is straightforward. The collected current is

$$I_i \simeq (en_i V_0 \sin\alpha + 0.37 en_i c_s)A, \qquad (5.83)$$

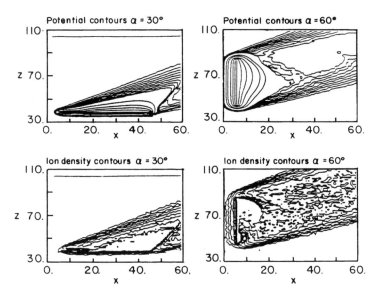

Figure 5.18. Near-Wake Region of a Biased Plate in a Plasma Flow at 30-deg and 60-deg Angle of Attack. [From Wang (1991).]

where A is the area of the ram surface. The first term arises from ions that would directly strike the surface. The second term arises from ions that are turned by the sheath into the surface. The second term is only important for angles of attack that are close to zero degrees. For large potentials where the sheath distance is large compared to the physical extent of the biased surface, the ions that reach the surface can come from a larger area than just the projected area onto the velocity vector. In general, this increase in the ion current must be determined numerically. It is conventional to use an ion focusing factor and write

$$I_i \simeq f(\Phi_w, L_t)(en_i V_0 \sin \alpha)A, \qquad (5.84)$$

where $f(\Phi_w, L_t)$ is given from particle-in-cell or particle-tracking calculations (Hastings and Cho, 1990). Typical values for the ion focusing factor range from three to eight.

The collection of electrons in a magnetic field to an attractive surface is a very complex problem. Considerable work has been done on collisionless, steady-state theories (Laframboise and Sonmor, 1993). (*Note:* Experimental data indicates that plasma turbulence and collisional ionization of neutrals may also play a significant role.) In the limit of large attractive potentials $\psi_w = -e\Phi_w/kT_e \gg 1$, the current collected to a spherical surface of radius r_p is (Laframboise and Sonmor, 1993)

$$i = \frac{I_e}{I_r} = \frac{1}{2} + \frac{2}{\sqrt{\pi}} \frac{\sqrt{\psi_w}}{\beta} + \frac{2}{\pi \beta^2}, \qquad (5.85)$$

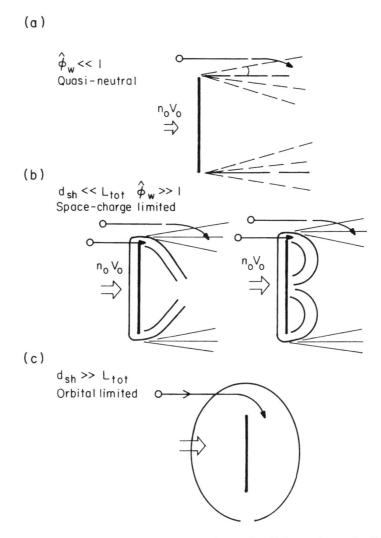

Figure 5.19. Transition of the Wake Structure from $e\Phi_w/kT_e \ll 1$ to $e\Phi_w/kT_e \gg 1$. [Taken from Wang (1991).]

where $I_r = 4\pi r_p^2 en_e \bar{c}_e/4$ is the random current to a sphere of radius r_p, and $\beta = r_p|\Omega_{p_e}|/\sqrt{2m_e/\pi kT_e}$ is the ratio of the probe radius to the mean attracted-particle gyroradius. The various terms in Eq. (5.85) are easy to interpret physically. In the limit of infinite magnetic field ($\beta \to \infty$), the electrons would be bound to the magnetic field with zero gyroradius. Hence, the only electrons that can get to the sphere are electrons that flow along the magnetic field. The projected area of the sphere on the magnetic field is πr_p^2, and electrons can flow parallel or antiparallel to the field, giving an effective collection area of $2\pi r_p^2$. Therefore, in this limit, $I_e/I_r = (2\pi r_p^2 en_e \bar{c}_e/4)/(4\pi r_p^2 en_e \bar{c}_e/4) = 1/2$. For a finite gyroradius, the electrons,

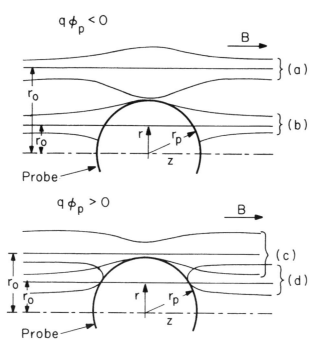

Figure 5.20. Magnetic Bottles for Attractive Potentials. From Laframboise and Sonmor, *Journal of Geophysical Research*, vol. 98, pp. 337–57, 1993, Copyright American Geophysical Union.

as they approach the attractive surface, gain energy and therefore their gyroradius increases. Hence, some of the electrons that are on field lines that do not pass through the probe will have a gyroradius large enough at the probe position that they strike the probe. This is illustrated in Figure 5.20 where the locus of the extrema of typical electron trajectories is shown. These are called magnetic bottles. This explanation accounts for the second term in Eq. (5.85). The third term arises from finite gyroradius orbits which completely encircle the axis of the probe. The first two terms of Eq. (5.85) are known as the Parker–Murphy limit (Parker and Murphy, 1967).

Experimental results from rocket experiments indicate that current collection in LEO can at times substantially exceed the results given by Eq. (5.85). Since large positive potentials are typically produced by electron beams emitted from the spacecraft, the interaction of the electron beam with the ambient environment, which can induce large return currents, is an important concern. This occurs in two ways. First the electron beam can induce plasma turbulence which acts as an effective source of collisions for the electrons (Linson, 1969; Hastings, 1986; Papadopoulos, 1986). These collisions can scatter the electrons sufficiently that all information about the magnetic field is lost to the electrons. The probe then collects as if there were no magnetic field present. The second way that an electron beam can produce enhanced return current is by beam-induced ionization of ambient neutrals (Neubert et al., 1990). This effectively increases the ambient electron density so that there is a larger pool of electrons from which to collect.

For repelled electrons, the electron distribution function will be Maxwellian and, even in the presence of a magnetic field, the current density to a surface will be

$$j_e = en_e(\bar{c}_e/4)\exp(e\Phi_w/kT_e) \quad \Phi_w < 0. \tag{5.86}$$

5.3.3 Spacecraft Potentials at LEO and Polar Orbits

At equilibrium, a spacecraft in LEO and polar orbit must collect zero net current. For a LEO spacecraft, the ambient electron current densities are of the order of milliamperes per square meter. Because this is much larger than the photoelectron current densities, the photoelectric current can be neglected. Since the mean ambient electron energies are low (0.1 to 0.2 eV), the electrons do not have enough energy to generate significant secondary or backscattered electron currents. Therefore, the major current densities to a LEO spacecraft are the ambient ion and electron current densities. Even with the ram enhancement, the ion current at zero potential is less than the ambient electron current. The spacecraft, therefore, must float negatively in order for the currents to balance. From Eqs. (5.83) and (5.86), the surface potential is the solution of

$$I_i - I_e \simeq Aen_e[V_0\sin\alpha - \bar{c}_e/4\exp(e\Phi_w/kT_e)] = 0. \tag{5.87}$$

This gives

$$\frac{e\Phi_w}{kT_e} = \ln(4V_0\sin\alpha/\bar{c}_e). \tag{5.88}$$

For $\alpha = \pi/2$, $T_e = 0.2$ eV, and $V_0 = 8$ km/s, this gives $\Phi_w = -0.45$ V. Hence, at LEO, the spacecraft surface potential is typically very small. A point to note from Eq. (5.88) is that the surface potential will vary naturally around the spacecraft depending on the angle that the surfaces make with the ram.

For polar orbits, in addition to the cold dense plasma typical of LEO, a spacecraft can experience intense fluxes of energetic electrons (see Section 3.3.3). Peak electron fluxes at 800 km as high as 50 μA/m^2 have been reported with peak energies of over 9,000 eV (Shuman et al., 1981). If a spacecraft is in eclipse, then the cold electron current density can drop as low as 100 μA/m^2. In this case, the hot electrons can make a significant contribution to the total electron flux to the spacecraft. The hot electrons are also sufficiently energetic that secondary electron emission becomes important. However, the question of the actual secondary current from a spacecraft is greatly complicated by the strong geomagnetic field. If the field is parallel to a surface, then electrons emitted from the surface cannot escape but will reimpact the surface one gyroradius away. When the hot electron current is important, the spacecraft must charge to a significant negative value to repel the combination of cold and hot electrons. In studies of the Defense Meteorological Satellites (DMSP), Gussenhoven et al. (1985) found that charging over 100 V was likely to occur under two conditions:

1. The ambient plasma density dropped below 10^{10} m^{-3}.
2. The integral number flux for electrons greater than 14 keV was greater than 10^{12} electrons/m^2-s-sr.

For polar orbits, the charging and associated phenomena under some conditions can resemble the situation at GEO. As an example, a severe charging event observed by a DMSP satellite occurred during a localized dropout of plasma when the spacecraft potential reached -462 V (Yeh and Gussenhoven, 1987). For this case, the measured parameters were as follows:

- Thermal ion density, 1.2×10^7 m^{-3}
- Integrated flux of electrons, 2.39×10^{13} electrons/m^2-s-sr
- Integrated flux of electrons greater then 14 keV, 2.33×10^{13} electrons/m^2-s-sr
- Integrated flux of ions, 1.48×10^{12} ions/m^2-s-sr

GEO charging codes are not adequate for LEO potential calculations. Since the plasma density at LEO is much larger than at GEO, it is too computationally expensive to determine the plasma density over large regions by integrating over all the particles, as is done in a code such as NASCAP. Therefore, in the solution of the Poisson equation, a semianalytical fit is often used for the charge density. For small potentials, the charge density $\rho \sim 1 + e\phi/kT$ (the orbit limit) whereas for large potentials, the charge density $\rho \sim 1/\sqrt{e\phi/kT}$ (the space-charge limit). Examples of charging codes that have been developed for LEO and PEO are NASCAP/LEO and POLAR. POLAR differs from NASCAP/LEO in that it includes the effect of the high-energy auroral streams often found in polar orbits. Both codes start with an assumed sheath boundary and potential distribution around the spacecraft. Particles are then tracked from the sheath boundary to the spacecraft and the particle trajectories are iterated with the potential structure until convergence is achieved. Although difficulties still exist, NASCAP/LEO, POLAR, and similar codes have allowed significant improvements in the understanding of how the low-altitude plasmas interact with orbiting spacecraft.

5.3.4 Particle-Beam Effects on Spacecraft Potentials

Electron beams have been used as particle emission sources on several spacecraft including the *Shuttle*. They are typically used as probes of the ambient environment

(Winckler, 1980; Neubert and Banks, 1992). When a beam current I_b is emitted, the spacecraft must rise to a positive potential large enough to either attract a neutralizing return current from the ambient electrons or large enough to attract back all of the emitted beam electrons. In the early days of rocket experiments, it was believed that a satellite would have to rise to a large, positive potential given by equating Eq. (5.85) to I_b. Experiments showed that the actual positive potential was, at most, a few hundred volts (Winckler, 1980). This discrepancy has been attributed to the low altitude of the rocket experiments (mostly below 200 km) and the initial conditions (the experiments were performed a few minutes after the rocket was exposed to vacuum while it was surrounded by a high-density ambient neutral environment and was still outgassing). The body potential need only rise to the level necessary to ionize the surrounding neutral gas (several times the typical ionization energy of 15 eV) to provide an ample supply of cold electrons to flow back to the surface. By contrast, on an experiment undertaken on the *Shuttle* (Sasaki et al., 1986), the *Shuttle* was observed to charge positively to the beam energy (5 kV) for altitudes above 200 km and when the conductive engine bells were in the wake of the *Shuttle*. In this case, the engine bells could not draw any return current from the highly depleted wake region and the *Shuttle* had to charge to the beam voltage to attract the beam electrons. As further confirmation, it was found that the ejection of low-energy plasma (Sasaki et al., 1987) dropped the Orbiter potential to near zero for as long as the low-energy plasma was present in the vicinity of the *Shuttle*. These experiments thus emphasize the importance of the ambient environment on beam operations.

5.3.5 *Potential Distribution on LEO Solar Arrays*

In Section 5.3.3, it was shown that the surface potential expected on LEO spacecraft is very small. However, if the spacecraft uses solar arrays to generate power, then the potential distribution on the solar arrays (at equilibrium) must be determined by the condition that no net current flow from the environment to the arrays and any surface electrically connected to the arrays and to space. Since the potential drop on the array is given by the number of cells, the system adjusts to provide zero net current by allowing one part of the array or spacecraft to collect electrons (i.e., positive with respect to space) and other parts of the array or spacecraft to collect ions (i.e., negative with respect to space).

The current collection to the array will occur at any location on the array where conductors or semiconductors are exposed to the space plasma. These include the metallic interconnectors between cells, which are usually bare to space as well as the edges of the solar cells. The latter have exposed cell material, which is a semiconductor. Because, as argued above, some part of the array must be positive with respect to space and other parts must be negative, there is a choice as to where to electrically connect the rest of the spacecraft to the array. The normal engineering procedure for U.S. spacecraft is to connect the spacecraft to the negative end of the array. This is called negative grounding. For Russian spacecraft the normal procedure is to connect to the point on the array where the potential is zero. This is called a floating ground. The other alternative is to connect to the positive end of the array. This choice is called a positive ground.

For negative and positive grounding, the potential of the conductive structure of the spacecraft will be negative or positive. For the space station, which has been designed as a negatively grounded structure, the structure potential was calculated in Hastings, Cho and Wang (1992). For the *space station*, the total current to the arrays and structure is

$$I = I_s + I_{s_r} + N_w N_p (I_{an} + I_{ap}) = 0, \tag{5.89}$$

where N_p is the number of solar-cell strings per wing, N_w is the number of wings on the station, I_{an} and I_{ap} are the ambient currents to the negative (n) and positive parts (p) of an array, I_{s_r} is the current collected by the conductive radiators, and I_s is the current collected by the rest of the structure. If there are N sites on each panel that can collect current, then the sites and associated areas can be ordered as $[A_0, A_1, A_2, \ldots, A_N]$ such that for sites A_0 to A_m the associated potentials satisfy $\phi_0 < \phi_1 < \phi_2 \ldots < \phi_m \leq 0$, and for sites A_{m+1} to A_N we have $0 < \phi_{m+1} < \phi_{m+2} \ldots < \phi_N$. Hence there are m sites with a potential that is negative or zero and $N - m$ sites with potential that is positive. If it is assumed that each string on each array wing will have solar cells each of which generates a voltage $\Delta\phi_c$, then

$$\phi_j - \phi_{j-1} = \Delta\phi_c \quad 1 < j < N. \tag{5.90}$$

In Eq. (5.89), the total current is set to zero (at equilibrium) to give the potential distribution subject to the constraint imposed by Eq. (5.90). If all of the areas are taken as $A_j = A$, then the total collecting area of the arrays is $A_a = N_w N_p N A$. With the normalization current given by $I_n = e n_e (\bar{c}_e / 4) A_a$ and the use of Eq. (5.90), for all regions,

$$\frac{I}{I_n}(m\Delta\phi_c, \alpha; T_i, T_e, M_0, A_{ex}/A_a, A_r/A_a) = 0. \tag{5.91}$$

Equation (5.91) can be solved for m and ϕ_0 as a function of angle of attack and parametrically as a function of T_i, T_e, M_0, the exposed area ratio A_{ex}/A_a (this is the insulator area exposed by oxygen erosion or punctures), and the radiator array ratio A_r/A_a.

To evaluate the structure potential for the *space station*, the solar array parameters are taken as $A = 6.5 \times 10^{-5}$ m^{-2} (this corresponds to the current collection area of

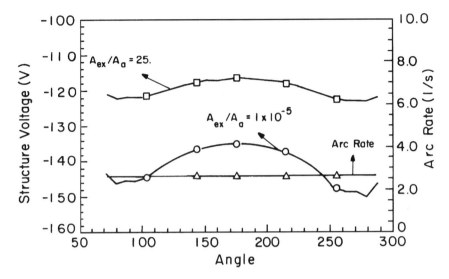

Figure 5.21. Structure Potential (V) and Arc Rate (1/s) Relative to Angle; $A_{ex}/A_a = 10^{-5}, 25$. [Taken from Hastings, Cho, and Wang (1992).]

the 8-mil by 8-cm gap along each edge of a solar cell), $N = 400$, $N_p = 82$, $N_w = 6$, and $\Delta\phi_c = 0.4$ V. This gives an array collection area of $A_a = 12.8$ m^2. The ambient Mach number of the ionic oxygen is $M_0 = 8$. The ambient ions have directed energy $\mathcal{E}_0 = 5$ eV. The length of the solar array panel in the flow direction is taken as 20 m, which for average ionospheric conditions gives the ratio $\lambda_D/L = 5 \times 10^{-4}$. The current-collecting area of the radiators is $A_r = 1,100$ m^2. The total area of the structure for the *space station* is taken as 2,500 m^2, giving $A_{ex}/A_a = 25$ for 13 percent of the insulator punctured or eroded.

Figure 5.21 is a plot of the structure potential for these parameters around the orbit measured from local midnight ($\theta = 0°$). For the case where there is no or very little exposed area on the structure (other than the radiators), the structure potential is quite negative. It is most negative near the dusk terminator ($\theta \simeq 270°$) because the array is in the wake and the structure must float as negatively as possible to attract ions. It is least negative at local noon ($\theta \simeq 180°$). Here, the radiators are normal to the flow and collect a large number of ions so that only a small additional ion current to the array is necessary to balance the electron current to the array. Although the structure potential for 13 percent of the area exposed is qualitatively similar to the case for no area exposed, the magnitude of the potential is everywhere reduced. This is because the increased ion collection of the structure requires a smaller ion collection by the array and therefore a smaller maximum potential.

For the positively biased parts of a solar array in LEO, it has been observed that the interconnects collect anomalously large electron currents above a critical voltage. This has been termed the snapover voltage from the sudden step increase in the current voltage curve [see, e.g., Thiemann and Bogus (1986)]. The experimental results for solar cells in a plasma chamber with a simulated LEO environment show that, for low positive voltages on the array, the current collected by a metallic solar-cell interconnector follow the classical Langmuir probe characteristic. Above 200 V, the current collected by the interconnectors becomes about an order of magnitude larger than expected from normal current collection. Potential profile measurements over the solar cells show the formation of potential barriers. Hysteresis effects are also observed in that the current collected by the interconnects shows a dependence on the time history of the voltage bias applied to the interconnects. There is general agreement that the anomalous current collection is due to the emission of secondary electrons from the dielectric surfaces which surround the conductors on the solar array (Hastings and Chang, 1989). The basic concept, as discussed earlier in this chapter, is that the dielectric surface, which must be in current balance, can have more than one solution for its surface voltage. If the dielectric has a high enough secondary electron emission coefficient, then it can achieve current balance in two ways: first, by repelling most of the incoming electrons so that the balance is between incoming ions and electrons; or second, by emitting secondary electrons so that the balance is between incoming primary and outgoing secondary electrons. Because the nearby conductor is always at a positive potential relative to the dielectric, secondary electrons that are emitted by the dielectric are collected by the conductor. The snapover is the jump of the surface potential and the associated electron current from one means of achieving current balance to another. The major system impact of the snapover phenomenon is the possibility of high Joule dissipation in the high-positive-voltage regions of a solar array. This heat load has to be determined for the operation of high-voltage solar arrays.

5.4 Spacecraft Arcing

5.4.1 Arcing on High-Voltage LEO and Polar Spacecraft

Conventionally, most solar array systems for U.S. space vehicles have a 28-V bus voltage. Future solar arrays, however, are being designed for much higher voltages to meet high power demands at low currents. For example, the *space station* has 160-V arrays. High voltages are more desirable than high currents for attaining high power levels because this choice minimizes the resistive loss and the mass of the cables in the power distribution system.

For the operation of high-voltage solar arrays in LEO, the high voltage can lead to arcing in two different ways. The simplest way is if the structure is negatively grounded. In that case, any thin insulator surface on the structure may suffer dielectric breakdown. The maximum arc rate for dielectric breakdown can be calculated as the inverse of the time to charge the insulator with thickness d to the electric field E_d, which is the dielectric strength. The rate is given by

$$R_{DB} = \frac{j_{\mathrm{ram}}}{\epsilon_d E_d} \qquad \frac{\phi_s}{d} > E_d$$
$$= 0 \qquad \frac{\phi_s}{d} < E_d. \qquad (5.92)$$

This assumes that once a dielectric breakdown occurs the surface recovers without any physical damage so that it can arc again. In fact, intense arcs will probably damage the surface to the point that the insulator material is removed. Therefore, the arc rate in Eq. (5.92) can be regarded as an upper bound on the true arc rate. This rate can be calculated for the *space station* which has a thin layer of Al_2O_3 over the aluminum structure. For the calculations, the extreme set of parameters for Al_2O_3 from Weast (1984) have been used, as well as the structure potential given in Figure 5.21. The set of parameters chosen to evaluate the maximum arc rate is $n_i = 10^{12}$ m^{-3}, $\epsilon_d/\epsilon_0 = 8.4$, $E_d = 6.3 \times 10^6$ V/m, and a dielectric thickness of $d = 5$ μm. Figure 5.21 is a graph of the maximum arc rate for the two cases considered. The maximum possible arc rate is independent of angle since the maximum value of the ram current is always assumed. For both cases, the structure will suffer an arc at a location several times a second. If the arcs are large, then this could potentially damage the thermal coating of the structure and lead to a loss of thermal control of the system.

The other type of arcing associated with high-voltage solar arrays is on the array itself. For negative biases below a voltage threshold (Thiemann and Bogus, 1986), approximately -200 V, arc discharges can occur. Experimentally, an arc discharge on a solar cell has been defined as a sudden current pulse up to the order of one ampere and lasting a few microseconds or less. Arcing can cause electromagnetic interference with instruments and damage to the solar cells (Thiemann, Schunk, and Bogus, 1990). Thus, there is a design trade-off between high voltages and damage from arcing unless solar cells can be designed to mitigate or even eliminate this arcing phenomenon.

Jongeward et al. (1985) proposed that there is a thin layer of insulating contaminant over the exposed interconnectors. Ions are attracted by the negative potential of the interconnectors and accumulate on the surface, resulting in the buildup of the electric field in the contaminant layer. The field causes electron emission into space, leading to subsequent heating and ionization in the layer. This positive-feedback mechanism eventually culminates in an arc discharge. Hastings, Weyl, and Kaufman (1990) proposed that neutral gas molecules are desorbed from the sides of the coverglass by the prebreakdown current, which has been observed experimentally. The neutral molecules build up over the interconnectors. Inside this surface gas layer, arcing occurs as a flashover discharge. Subsequent work by Cho, Hastings, and Kuninaka (Cho and Hastings, 1991; Hastings, Cho, and Kuninaka, 1992) combined ideas from these two theories to analyze the charging of the region near the plasma, dielectric, and conductor called the triple junction.

Cho and Hastings determined the following from theoretical and numerical work:

(1) Ambient ions charge the coverglass front surface but leave the side surface of the coverglass effectively uncharged.
(2) Ambient ions cause electron emission from the interconnector. These electrons charge the side surface through secondary electron emission from the side surface to a steady state unless enhanced electric-field electron emission (EFEE) becomes significant.
(3) EFEE will charge the side surface if there is an electron emission site close to the triple junction with a high field enhancement factor, β. The field enhancement factor measures how much the local electric field is increased over the average field at the electron emission site. A value of $\beta = 1$ implies no enhancement, whereas $\beta = 100$ implies that the local field is concentrated by two orders of magnitude relative to the average field. Values of $\beta > 1$ are always associated with surface imperfections (Latham, 1993). Typically, values of $\beta < 100$ are associated with metallic microwhiskers, whereas values of $\beta > 100$ are associated with dielectric inclusions.
(4) EFEE can result in collisional ionization of neutrals desorbed from the coverglass; this is the arc discharge.
(5) The arc rate is the inverse of the sum of the ion(τ_{ion}) and EFEE (τ_{EFEE}) charging times.
(6) For high voltages and high β values, the arc rate is mainly determined by the ion charging time, and for low voltages and low β values, the arc rate is dominated by the EFEE charging time.

In practice, the arcing rate is calculated by dividing the number of arcs during a finite experiment time by the experiment time. The time between arcs, τ_{arc}, is the shortest charging time, $\tau_{\text{EFEE}} + \tau_{\text{ion}}$, among all the emission sites on the interconnector surface.

Two flight experiments (Grier, 1983) have specifically addressed this issue. Both were piggybacked on Delta satellite launches, which inserted them into 800-km polar orbits. In both experiments, a small set of solar cells was biased to a high positive and negative voltage with respect to the space environment. The first experiment, PIX-I, was launched in March 1978, and confirmed that arcing was inherent to the space plasma for standard-design solar cells. The second experiment, PIX-II, was launched in January 1983. The results were limited by primitive diagnostics as well as an uncontrolled tumbling of the Delta second stage. The results are shown in Figure 5.22 along with the PIX-II ground experiments as well as data from two other ground-based experiments.

Hastings et al. (1992) calculated the arc rate numerically for the PIX-II flight and ground experiments as well as the space flyer unit (SFU) ground experiments. The

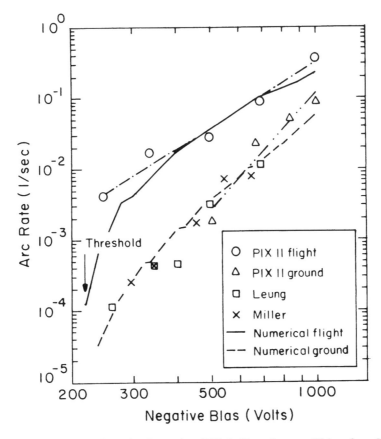

Figure 5.22. Experimental Data for Ground and Flight Experiments. [Taken from Hastings, Cho, and Kuninaka (1992).]

results show good agreement with the data over the range that the data exist. This agreement and the understanding of the processes governing the arcing allows the assessment of mitigation strategies and design modifications to conventional solar cells. Specifically, the model suggests that the following strategies will decrease the arc rate: (1) increasing the interconnector work function; (2) increasing the dielectric thickness; (3) decreasing the secondary electron yield to near or below one; (4) decreasing the ratio $\epsilon_{d1}/\epsilon_{d2}$, especially to below one; (5) overhanging the coverglass, particularly past the critical overhang that causes electrons to be trapped on the coverglass back surface. Increasing the work function affects τ_{efee} significantly but only affects the arc rate for extreme values. The dielectric thickness has a slightly larger effect on the arc rate, particularly for low thicknesses and low voltages. The secondary electron parameters that eliminate the electric-field runaway eliminate arcing. The best results for decreasing the arc rate are achieved by changing the dielectric constants and by extending the coverglass over the interconnector.

5.5 Electrodynamic Tethers

A new and novel concept introduced in the 1980s was that of the space tether. Tethers in space can be used for a variety of applications such as power generation, propulsion, remote atmospheric sensing, momentum transfer for orbital maneuvers, microgravity experimentation, and artificial gravity generation. In general, a tether is a long cable (up to 100 km or longer) that connects two or more spacecraft or scientific packages. Here, the discussion focuses on the electrodynamic tether. Electrodynamic tethers are conducting wires that can be either insulated (in part or in whole) or bare, and that make use of an ambient magnetic field to induce a voltage drop across their length.

Consider a tether in LEO as in Figure 5.23. For prograde, low-inclination orbits, the velocity vector ($\vec{v}$) of a vertically oriented tether points eastward, and is almost perpendicular to the magnetic-field lines ($\vec{B}$), which run south to north. Charged particles in the tether will experience a force given by the Lorentz relation [see Eq. (2.42)]. Hence the motional electric field induces an electromotive force (EMF) given by

$$\Phi_{\text{induced}} = \int \vec{v} \times \vec{B} \cdot d\vec{l}, \tag{5.93}$$

where $d\vec{l}$ is an element of length along the tether. The velocity $\vec{v}$ is the relative velocity of the tether to the ambient plasma that is corotating with the Earth. In most cases, the tether is assumed to be straight, and so, the induced voltage becomes

$$\Phi_{\text{induced}} = \vec{v} \times \vec{B} \cdot \vec{L}, \tag{5.94}$$

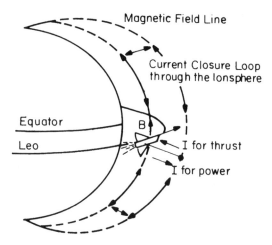

Figure 5.23. Tether Schematic in LEO. Taken from SamantaRoy and Hastings (1993). Reprinted by Permission of Kluwer Academic Publishers.

where L is the length of the tether. If current is allowed to flow through the tether, the electromagnetic force on a rigid tether will be

$$\vec{F}_{EM} = L\vec{I} \times \vec{B}. \tag{5.95}$$

The direction of the current will determine whether the direction of the force will be a thrusting force or a drag force. For an upward- (or downward-) deployed tether, if the current flows up (away from the Earth), power is generated at the cost of orbital energy because the electromagnetic force is antiparallel to the direction of motion. Conversely, if a power supply large enough to reverse the induced EMF drives the current down, the electromagnetic-force vector acts parallel to the direction of motion and produces thrust. Of course, these directions are reversed for retrograde orbits. In an electrodynamic tether, the current that flows is actually the electrons that are collected from the ambient ionospheric plasma. Electrons are drawn in on one end of the tether and are ejected out the other, the particular end depending on whether the tether is thrusting or generating power. The end of the tether that must collect electrons and/or eject ions is called the anode. The other end that ejects electrons and/or collects ions is called the cathode. Paramount to this process of current flow are two issues. One is the ionospheric resistance to a current flowing through it. The other is the actual ability of the anode end of the tether to collect the electrons with little voltage loss. The electron-emitting end does not seem to pose a large problem because space tests have demonstrated that large currents can be ejected with little voltage drop. The ionospheric impedance is due to a complicated electromagnetic phenomenon. Analogous to a ship creating waves as it moves through the water, a moving conductor through a plasma generates electromagnetic waves. These waves dissipate energy with an effective resistance, called the radiation impedance, which has been examined in detail (Belcastro, Veltri, and Dobrownoly, 1982; Barnett and Olbert, 1986; Hastings and Wang, 1987; Donahue et al., 1992). It has been found that radiation is emitted in three distinct bands: the Alfvén, the lower hybrid, and the upper hybrid. For a long tether, the Alfvén band is the most significant and the impedance is highly dependent on the ambient ionospheric conditions, that is, the electron density, which varies considerably over an orbit. The impedance is also a strong function of the dimensions of the collecting ends of the tether system: the impedance is inversely proportional to the diameter. Although there is controversy over the impedances that will be encountered by long tether systems, it is strongly believed that the order of magnitude will be a few ohms.

Several different options exist for electron collection, such as a passive large surface (like a balloon), a passive grid, a plasma contactor, or a light ion emitter. The most promising of these devices is the plasma contactor (see Section 5.6.1). The ambient electron density in LEO is rather low. Collecting the required current requires a very large surface area (100–$1,000$ m^2 for 1 A). Instead of a large physical area, plasma contactors create a plasma cloud that expands and collects ambient electrons while emitting ions. Note that these contactors operate by ejecting fully or partially ionized gas. For a device using argon, the mass flow rate is about 13 kg $\cdot$ yr^{-1} A^{-1}. A recently proposed method for electron collection is to leave part of the tether bare (i.e., to have only part or the whole of the conducting wire uninsulated). For a 20-km tether, up to 10 km, which represents a rather large area, will be positive to the plasma and can collect electrons. The inherent advantage of this scheme is the absence of the mass and complexity of a collecting contactor.

In a strict sense, the picture of electron emission and collection into the ionosphere as a DC phenomenon is not entirely correct. Electrons emitted and collected are constrained to travel along the magnetic-field lines (or flux tubes), which can be thought of as parallel transmission lines. These transmission lines are excited as the tether ends contact them so that the phenomenon is fundamentally AC. However, because the magnetic-field lines in reality form a continuous medium, the current flow is DC. A circuit equation can be written consisting of the various voltage drops for the tether system. For a tether generating power,

$$\Phi_{\text{induced}} = \Delta V_A + \Delta V_C + IZ_T + IR_T + IR_L, \tag{5.96}$$

where ΔV_A and ΔV_C are the voltage drops across the anode and cathode, respectively, Z_T is an effective ionospheric impedance, R_T is the tether ohmic resistance, and R_L is the load. If an efficiency, η, is defined as

$$\eta = \frac{Power_{\text{load}}}{Power_{\text{total}}} = \frac{I V_L}{I \Phi_{\text{induced}}} = \frac{V_L}{\Phi_{\text{induced}}}, \tag{5.97}$$

then the circuit equation can be rewritten as

$$\Delta V_A + \Delta V_C + IZ_T + IR_T = \Phi_{\text{induced}}(1 - \eta). \tag{5.98}$$

It can be shown that for any given operating conditions (i.e., Φ_{induced} and electron density), there exists a unique value of η where the power generated, $I^2 R_L$, is maximized. A similar equation can be written for a tether generating thrust, except now an onboard power supply is required to reverse the current,

$$\Phi_{\text{induced}} + \Delta V_A + \Delta V_C + IZ_T + IR_T = V_{\text{PS}}, \tag{5.99}$$

where V_{PS} is the voltage of the power supply. Figure 5.24 contains schematics of these tether circuits.

Over the past several years, there have been a number of system or engineering studies conducted that concentrated on the electrodynamic aspects of tethers. These studies have examined the uses of tethers in propulsion and/or power-generation

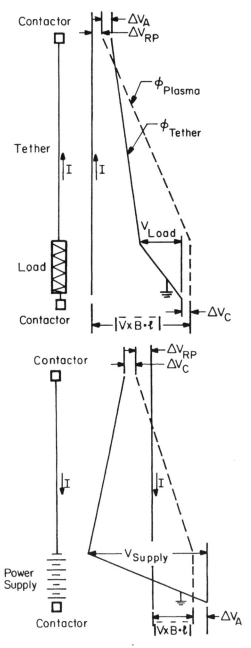

Figure 5.24. Tether Circuit for Power and Thrust Generation. The Total Potential Drop is $|\vec{V} \times \vec{B} \cdot \vec{L}|$. This Appears Across the Cathode ΔV_c, Across the Load V_{load}, Along the Tether ϕ_{tether}, and Across the Anode ΔV_A and Through the Plasma ΔV_{RP}. Outside the Tether, the Local Plasma Potential is ϕ_{plasma}. Used with Permission from SamantaRoy and Hastings (1993). Reprinted by Permission of Kluwer Academic Publishers.

applications (Martinez-Sanchez and Hastings, 1987; SamantaRoy, Hastings, and Ahedo, 1992). They found that, because of magnetic-field variations, the power generated varied by as much as ± 20 percent, and thus batteries were required for power-leveling purposes. The concept of orbital energy storage, or mixed-mode operation, was also introduced. The idea was that a tether would thrust in the day, drawing power from solar arrays, and generate power during the night, the power/thrust levels being adjusted so that the semimajor axis (i.e., the energy) of the orbit would remain constant. These studies used detailed computer models to simulate the performance of a tether in LEO, and included a highly accurate model of the geomagnetic field, the latest IRI model, realistic orbital dynamics, and temperature effects. In addition, the electron collection performance of a contactor and a bare-wire tether, both separately and in combination, was compared and contrasted. The power and thrust generated by a bare-wire tether had a higher dependence on the geomagnetic and ionospheric fluctuations; however, depending on the performance of the contactor, the combination of a bare tether and contactor can substantially boost performance for power generation. As a pure thruster, the contactor tether was examined at constant current, voltage, thrust, and power, and it was found that the best mode of operation was with constant power, with resulting power/thrust ratios better than those for ion or MPD engines. For power generation, geomagnetic variability was still a major difficulty, as observed in previous studies, but a control strategy was developed to greatly reduce the impact of this highly undesirable condition. In addition, operation at equatorial orbits was much more beneficial for system performance. It was concluded that tethers offered greater potential than previously envisioned.

5.6 Plasma Sources on Spacecraft

5.6.1 Plasma Contactors

Plasma contactors are devices for generating plasma clouds which allow the passage of charge between an electrode and an ambient plasma. They have been proposed for use in power-generating devices such as electrodynamic tethers (Martinez-Sanchez and Hastings, 1987) because they may substantially reduce the impedance of the electron-current collection from the ionosphere and make the emission of electrons much less energetically expensive than using an electron gun. At GEO, they also have been used to reduce differential surface potentials by flooding the surfaces with plasma and "shorting" the surface potentials. Most recent research, however, has concentrated on plasma contactors used at an anode to collect electrons in the ionosphere or other ambient plasma. Such a contactor will emit ions as well as collect electrons. Two figures of merit for such a contactor are its impedance, ϕ_0 / I, and the gain, ξ, defined as

$$\xi = I/I_i(r_{\text{anode}}).$$

For a contactor used with a tether, the impedance determines the maximum power that can be generated because the total tether potential ϕ_{induced} is fixed. The gain is important because it determines the rate at which gas must be expended (to produce ions) for a given total current. If the gain is high, less gas is used to collect a given current.

Both the impedance and the gain depend on the current. In general, there is a trade-off: at very low current, both high gain and low impedance are possible, but the power is low. At high current, high gain can be obtained only at the cost of very high impedance (again resulting in low power). Low impedance and high power are possible only with low gain. To illustrate these trends, consider the extreme limits. When the current is equal to the electron saturation current of the ambient plasma over the surface area of the physical anode, then the gain is infinite (because no ions need be emitted to draw this much electron current), and the contactor impedance is zero. The power, however, for LEO and practical tether and anode parameters is, at most, tens of watts. Arbitrarily large current (and high power) may be obtained by emitting a large ion current, but unless the anode potential is high enough, it will not be possible to collect many electrons across the magnetic field, and the gain will approach unity.

A plasma contactor cloud will consist of several different regions (Hastings and Gatsonis, 1988; Szuszczewicz, 1986). The first is an inner core where the cloud will be isotropic because the two major directions of anisotropy, namely the Earth's magnetic field and the direction of motion of the source will be shielded by the dense plasma from the contactor source. There will then be two outer regions where the two directions of anisotropy are manifested. Data and theory support the idea that a substantial current of ambient electrons can be collected only from field lines that pass through the inner-core region (Hastings and Gatsonis, 1988; Hastings and Blandino, 1989).

There has been much debate about the size of the core region from which electrons can be collected. One estimate is obtained by matching the cloud density to the ambient density (Parks, Mandell, and Katz, 1982; Parks and Katz, 1987),

$$n_{\text{cloud}}(r_{\text{core}}) \approx n_{\text{ea}}$$

and another by taking magnetic-field effects into account (Hastings, 1987),

$$\nu_e(r_{\text{core}}) \approx \omega_{\text{ce}},$$

where ν_e is the radially dependent electron collision frequency (including effective "collisions" due to turbulence). A third estimate is obtained by requiring regularity of the self-consistent potential (Iess and Dobrownoly, 1989)

$$\frac{\partial \phi}{\partial r}\bigg|_{r_{\text{core}}} \approx 0.$$

A fourth estimate comes by requiring a consistent space-charge-limited flow inside the core (Wei and Wilbur, 1986):

$$m_i n_i u_i^2 \big|_{r_{\text{core}}} \approx m_e n_e u_e^2 \big|_{r_{\text{core}}}.$$

These diverse theories give a wide range of current enhancement factors for the plasma cloud.

If a core cloud of radius r_{core} is assumed, then continuity of current gives

$$I = I_i(r_{\text{anode}}) + I_e(r_{\text{anode}}) = I_i(r_{\text{core}}) + I_e(r_{\text{core}}),$$

and the gain is

$$\xi = \frac{I_e(r_{\text{core}})}{I_i(r_{\text{anode}})} + \frac{I_i(r_{\text{core}}) - I_i(r_{\text{anode}})}{I_i(r_{\text{anode}})} + 1.$$

Plasma contactor clouds enhance or produce electron current flow through two possible paths. First (the first term on the right-hand side of the equation), they can serve as virtual anodes through which electrons from far away can be drawn and collected to the real anode at the center of the cloud. Second (the second term on the right-hand side), the neutral gas associated with the cloud can become ionized, creating electron–ion pairs. The electrons will be collected to the anode and the ions will be repelled. For an electrodynamic tether, however, ionization of contactor neutrals is not an efficient use of neutral gas; if this is the only means by which the current is enhanced, then the same neutral gas can be used more efficiently by ionizing it internally in an ion source. Plasma contactors will be useful if they enable the ionosphere to supply electrons. The two sources of electrons in the ionosphere are the ionospheric plasma and the ionospheric neutrals. The mean free path for ionization of the ionospheric neutral gas, however, is so long (many kilometers) that ionization of this gas on the length scale of the plasma contactor cloud is highly unlikely. Therefore, plasma contactors are useful with electrodynamic tethers only if they enhance current by collecting ambient electrons from the ionosphere. The collected electron current $I_e(r_{\text{core}})$ generally will be the saturation current times the area of the core cloud $4\pi r_{\text{core}}^2$, or, if the contactor is only collecting electrons along magnetic-field lines running into the core cloud, then $I_e(r_{\text{core}})$ will be the saturation current times $2\pi r_{\text{core}}^2$ (if the core cloud is not spherical but is elongated in the direction of the magnetic field, then r_{core} is the minor radius across the magnetic field). For this reason, the size of r_{core} is crucial to the effectiveness of plasma contactors as electron collectors in space.

Outside of the core region, a double layer will form that will mediate the interaction between the core and the ambient plasma (Gerver, Hastings, and Oberhardt, 1990; Dobrownoly and Melchioni, 1992, 1993). Experimental data indicate that this core cloud can be large. Indeed, contrary to the theoretical expectations given above, it seems to be independent of the magnetic field (Dobrownoly and Melchioni, 1993).

Plasma contactors also have been used for grounding of GEO spacecraft (Purvis and Bartlett, 1980). They operate as low-energy sources of cold plasma (Sasaki et al., 1987), which neutralizes the charge resident on the different surfaces of a spacecraft. It has been shown that the plasma cloud can discharge a biased spacecraft in milliseconds and keep the spacecraft grounded for as long as the cloud is resident in the vicinity of the spacecraft.

5.6.2 *Electric Propulsion Engines*

Another plasma source on a spacecraft arises when a plasma thruster is used for station keeping or for primary thrust. It is shown in Section 4.3.2 that thrusters that emit neutrals can lead to backflow onto the spacecraft and associated contamination. The same concerns exist for a plasma thruster. For example, in ion thruster plumes, a low-energy plasma is created by charge-exchange (CEX) processes and can expand around a spacecraft, leading to a current drain on high-voltage surfaces. The enhanced plasma density can also lead to attenuation and refraction of electromagnetic-wave transmission and reception. In addition, many thrusters emit heavy-metal species, both charged and uncharged, because of erosion, which can easily adhere to spacecraft surfaces.

In contrast to the case of a neutral gas, the backflow from a plasma thruster is very sensitive to local electric fields and to collective effects. Backflow contamination can lead to sputtering and effluent deposition that can affect such systems on the spacecraft as the solar arrays, thermal control surfaces, optical sensors, communications, science instrumentation, general structural properties of materials, and spacecraft charging. Ion thrusters have been studied the most because of their maturity and the relatively large database with which to compare results. One issue associated with ion thrusters is that complete ionization cannot be achieved with reasonable levels of power, and hence, neutral gas is emitted at thermal speeds. These slow neutrals are of interest because they charge-exchange with the fast beam ions, producing fast neutrals and slow ions, which can be influenced by local electric fields in the plume. The electric-field structure in the plume, as seen in experiments (Carruth and Brady, 1981) and in computational models, is radial, and hence the slow ions are pushed out of the beam and move back toward the spacecraft. Because most of the CEX ions are created within a few beam radii downstream of the thruster,

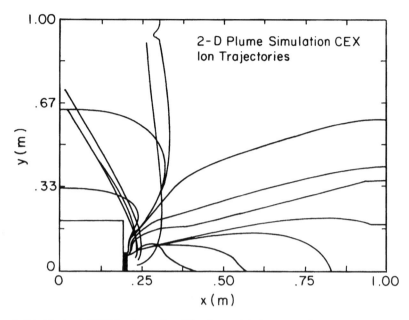

Figure 5.25. Typical CEX Trajectories. The Potential Contours Show the Beam Structure While the CEX Ions Born in Front of the Grid are Ejected Radially.

local electric fields near the thruster play an important role. This is illustrated in Figure 5.25. In this figure, the potential contours and some typical ions formed by CEX are shown for an ion engine. The main beam ions give rise to a local radial field near the edge of the beam. Any slow-moving CEX ion that gets there will be accelerated outward with some component of the velocity going back toward the spacecraft. This will also occur for any ions created by sputtering of grid material. This backflow must be modeled and taken into account in placement of the thrusters on a spacecraft.

6

The Space Radiation Environment

6.1 Introduction

As electronic components have grown smaller in size and power and have increased in complexity, their enhanced sensitivity to the space radiation environment and its effects has become a major source of concern for the spacecraft engineer. The three primary considerations in the design of spacecraft are the description of the sources of space radiation, the determination of how that radiation propagates through material, and, thirdly, how radiation affects specific circuit components. As the natural and man-made space radiation environments were introduced in Chapter 3, the objective of this chapter is to address the latter two aspects of the radiation problem. In particular, because the "ambient" environment is typically only relevant to the outer surface of a space vehicle, it is necessary to treat the propagation of the external environment through the complex surrounding structures to the point inside the spacecraft where knowledge of the internal radiation environment is required. Although it is not possible to treat in detail all aspects of the problem of the radiation environment within a spacecraft, by dividing the problem into three parts – external environment, propagation, and internal environment – a basis for understanding the process of protecting a spacecraft from radiation will be established that can be applied to a wide range of radiation problems.

6.2 Radiation Interactions with Matter

From the standpoint of radiation interactions with matter, three particle families need to be considered:

(1) photons (primarily EUV, X rays, and gamma rays),
(2) charged particles (protons, electrons, and heavy ions),
(3) neutrons.

Although numerous, more exotic particles (e.g., positrons, muons, mesons) exist, these three families account for the vast majority of interactions of concern to the spacecraft engineer. In addition, for the impacting particles, mass, charge, and kinetic energy are the principal physical characteristics of interest, whereas mass and density are the key characteristics for the target material. Here, the types of interactions are discussed in terms of these three particle families. The effects of the shielding on these particles is manifested in terms of energy deposited in a volume (dose) or the energy deposited per unit length in the target material (LET) after traversing a specified thickness of shielding (see Section 2.5.2). The radiation shielding calculation necessary to determine the environment inside a spacecraft thus breaks down into a three-step process for each particle:

(1) definition of the external radiation environment,
(2) propagation of that environment through the shield and calculation of the subsequent changes in the spectrum up to the target,
(3) estimation of the total energy and/or the energy deposition rate at the target.

The external radiation environment is detailed in Section 3.4. In this section, the latter two issues are addressed. An important factor that should be kept in mind when considering these two issues is the importance of the cascade process to the final result. In this process, one incident particle interacts to produce many secondary particles that may be very different from the incident particle (e.g., electrons may generate photons or vice versa). These secondaries, in turn, generate their own secondaries, leading to a complex mix of many different photons and particles. This process repeats until the point of interest is reached or until all of the initial particle energy is dissipated. Rather than address this process in its entirety, it is first broken down into the individual, distinct single-particle interactions. The final part of this section describes how, given the characteristics of these individual interactions, Monte Carlo techniques can be used to estimate the gross effects of the cascade process. In the simplest models, analytical expressions are fitted to these results or to actual measurements to give estimates of the end products produced by the cascade as a function of depth in the shielding material. The reality is that the cascade process is basically probabilistic and too random to be precisely modeled analytically. However, given that the analytical fits typically give adequate results in most cases of interest to the engineer, models based on them are the focus here for practical radiation shielding calculations.

6.2.1 Single-Particle Interactions

The study of the interactions of a single high-energy particle, such as a photon, neutron, or charged particle, with matter forms a major subdivision of the physical sciences. Rather than present a detailed quantitative review of each of these interactions, a qualitative description is presented for each of the main interactions. This is supplemented with a short quantitative discussion of the interaction where appropriate. The reader is referred to detailed quantitative reviews of each of the processes, such as can be found in Particle Data Group (1990). In most practical cases, however, the results presented here should suffice for actual computations.

6.2.1.1 Photon Interactions

Photons, which propagate at the speed of light and have no charge or rest mass, interact primarily through the photoelectric effect, Compton scattering, and pair production. All of these interactions generate free electrons. Consider first the photoelectron process, the probability of which decreases with increasing photon energy and increases with Z, the atomic number of the target nucleus. In the photoelectron process, the photon is completely absorbed by the emitted (typically) outer-shell electron of the target atom. In one case, however, subsequent interactions are possible, that is, if the photon is energetic enough to emit K-shell electrons (inner-shell electrons), then this process will dominate approximately 80 percent of the time over the emission of outer-shell electrons. When an L-shell (or outer-shell) electron subsequently drops down to fill the K-shell vacancy, it can emit either an additional X ray or a low-energy Auger electron from the L-shell (dependent on the Z of the target material). In Compton scattering, the incident photon is not completely absorbed because the photon is of much greater energy than the atomic-electron-binding energy of the target. Part of the photon energy goes to scattering the atomic electron (called a Compton electron) and the rest into a scattered, lower-energy photon. Pair production takes place for photons at energies of 1.02 MeV or higher. A photon of this energy is completely absorbed by a high-Z material. A positron–electron pair is then formed. Figure 6.1 compares the ranges over which each of the three interactions dominates as functions of Z and energy. For reference, in silicon, the photoelectron effect dominates at energies <50 keV, pair production takes place at energies >20 MeV, and Compton scattering occurs at intermediate energies. The products of these interactions (electrons, photons, and positrons) can, of course, further interact with the target material, producing a complex cascade of electrons and photons.

UV photons interacting with complex organic polymers can cause material degradation and change the properties of the materials [see review by Stiegman and Liang (1993)]. Extensive flight data from the *SCATHA* satellite indicates that the solar absorptance of thermal control coatings increases under UV irradiation. This is illustrated by the data in Figure 6.2 from *SCATHA*. The solar absorptance changed by more than a factor of two over a decade. To prove that the change in solar absorptance was due to UV rather than to contaminants (see Section 4.3), the data were compared with similar data from gold surfaces on *SCATHA*. The polished gold surfaces showed no change in absorptivity over the same period, proving that the

PHOTON INTERACTIONS

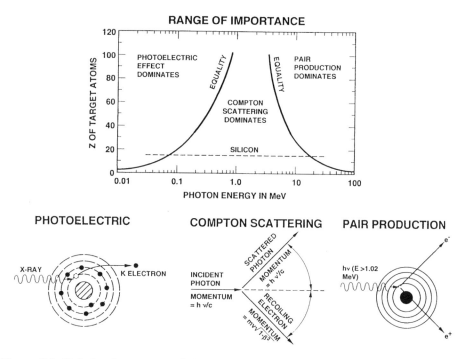

Figure 6.1. Relative Importance of the Three Photon Interactions as a Function of Z and Photon Energy. Solid Lines Correspond to Equal-Interaction Cross Sections for the Neighboring Effects. The Dashed Line Illustrates the Situation for Photon Interactions with Silicon.

effect is due to UV exposure. This change in absorptance is of considerable concern for spacecraft design because the temperature of a spacecraft is proportional to the solar absorptance, and too great a temperature increase can lead to loss of the spacecraft. The cause of this problem is the absorption of the UV photons by organic molecules. This absorption results in highly reactive excited states that can react on a molecular level and lead to macroscopic changes in the surface properties. Careful selection of radiation-resistant, external surface materials is thus an important consideration in spacecraft design.

6.2.1.2 *Charged-Particle Interactions*

Charged particles interact with matter primarily in two ways: Rutherford scattering and nuclear interactions. Rutherford (or Coulomb) scattering, in which the charged particle interacts with the electric field of the target atom, typically dominates. It results in both excitation and ionization of atomic electrons and can, for sufficiently

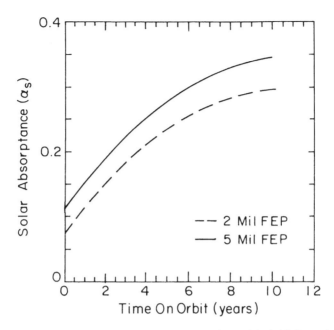

Figure 6.2. Changes in Solar Absorptivity of Silver Teflon with Orbit Duration. Used with the Permission of DeWitt et al. (1993). Reprinted by Permission of Kluwer Academic Publishers.

energetic impacts, transfer enough energy to displace atoms within the lattice structure. As an example, for electrons, a minimum energy of approximately 150 keV is required to cause displacement in silicon, whereas only 160 eV is required for protons. This can be explained in the following manner.

The maximum energy that can be given to an atom of mass M by an incoming particle of mass m and energy E is

$$\Delta E = 4E \frac{mM}{(M+m)^2}.$$ (6.1)

If the energy to displace the atom of mass M is E_{th}, from Eq. (6.1) the threshold energy for the incoming particle is

$$E_s = E_{th} \frac{(M+m)^2}{4mM}.$$ (6.2)

For silicon, $E_{th} = 12.9$ eV, which leads to the minimum energies given above for protons and electrons.

Nuclear interactions, where the impacting particle actually interacts with the atomic nucleus, can result in elastic or inelastic scattering and transmutation (through fusion or fission). As an example, a nucleus can absorb a proton and emit an alpha particle. This process, also called spallation, and the recoil atoms that result from

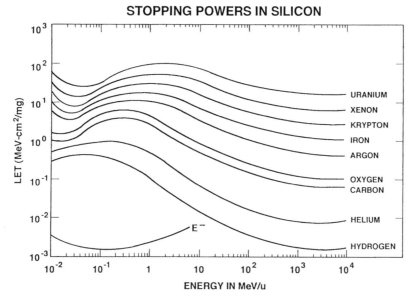

Figure 6.3. Stopping Power (or LET) in MeV-cm^2/mg Relative to Energy per Atomic Mass Unit for a Variety of Ions (MeV/μ) and Electrons (MeV) in Silicon.

displacement can transform a relatively benign proton environment into a heavy-ion environment capable of generating SEUs because the heavy ions have much larger LETs than do protons. Also, long-term exposure to the space radiation environment can, through transmutation, lead to making the spacecraft material itself radioactive.

One quantitative measure of the interaction of a high-energy particle with matter is stopping power or energy loss per unit length in a given material. Consider low-energy electrons (approximately 10 keV), which primarily cause ionization. The amount of energy deposited per unit length by the latter and by protons determines the amount of ionization and can be determined easily from stopping-power tables. Stopping power is also essential in calculating the Heinrich flux (necessary for most SEU calculations, see later). Stopping power (or LET) in terms of MeV-cm^2/mg is given in Figure 6.3 for electrons, protons, and heavy ions in silicon.

A second quantitative measure of high-energy particle interactions closely related to stopping power is the penetration depth/range or the maximum distance a particle of a given energy can penetrate. This depth is a rough estimate of the minimum cutoff energy for a given thickness of spacecraft shielding and hence its effectiveness. Figure 6.4 compares the penetration depth of electrons and protons in aluminum for different energies. Note in particular that an electron at 1 MeV penetrates over 100 times more shielding ($\approx$0.2 cm) than a 1-MeV proton ($\approx$0.0015 cm). Similarly, it takes a 20-MeV proton to penetrate the same depth as a 1-MeV electron. A typical shielding level is 0.1–0.2 cm (40–80 mils). Thus, it is common

CHARGED PARTICLE INTERACTIONS
PROTON/ELECTRON ENERGY vs PENETRATION DEPTH FOR AL

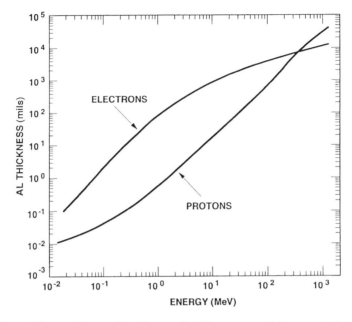

Figure 6.4. Minimum Penetration Energy for Electrons and Protons Relative to Shield Thickness.

to compare the integral dose for $E > 1$-MeV electrons with $E > 20$-MeV protons because these are the primary contributors to the radiation environment behind the spacecraft shield.

6.2.1.3 Neutron Interactions

Neutron interactions with matter can result in elastic scattering, inelastic scattering, and nuclear transmutation. In elastic scattering, the neutron is not captured but transfers some of its energy to the target atom, which can be displaced from its lattice position. It occurs only if the incident neutron has more energy than required for displacement – typically >25 eV. The target atom, referred to as a primary recoil or knock-on, can in turn cause ionization or further displacement damage. In inelastic scattering, the neutron capture by the nucleus is followed by the emission of a lower-energy neutron. The kinetic energy lost in the process can result in displacement or can excite the atomic nucleus, which returns to the ground state by emitting a gamma ray. In transmutation, the capture of the neutron can change the atomic isotope, cause fission, or cause the emission of another particle such as a proton or an alpha particle. For silicon, the dominant process is displacement and ionization for neutrons with energy 1 MeV or higher. The effects of neutron dose are typically not considered in radiation calculations except for the prompt nuclear environment, nuclear power sources, or for evaluating the results of ground testing.

6.3 Modeling the Effects of Shielding

Section 6.2 has briefly described the basic interactions between single particles and matter. If the detailed evolution of a particle passing through matter is followed, the interaction of the particle with the shielding becomes increasingly complex as each interaction gives rise to a cascade of byproducts. Fortunately, as each interaction disperses the energy into more byproducts, a point is reached where the byproducts and the original incident particle (if it still exists) no longer have sufficient energy to excite further interactions. Thus the process has a finite conclusion. Although it is common practice to use Monte Carlo techniques to model the detailed passage of a particle through shielding, the computer codes that accomplish this can require supercomputers or take many hours to perform the calculations. As a consequence, it is common to run the detailed codes only for a range of variables and then use this information to derive analytical fits to the end products of the multiple-particle interactions that are created following a single particle impact. The effects of these byproducts are then approximated in terms of displacement damage, energy deposition, or ionization (or electron-hole creation). It is normally these algorithms, not the detailed computations, that are used to actually model radiation effects.

As a specific example of the Monte Carlo results, consider electrons. Electrons are particularly easily scattered in a material. Rather than passing through the material, they and the secondary electrons they generate are scattered into the material. This behavior is illustrated in Figure 6.5 which is a computer simulation (Monte Carlo) of the trajectories of 5-MeV electrons impacting on "infinitely thick" aluminum and lead targets (Holmes-Siedle and Adams, 1993). Note that many of the electrons are actually scattered back out of the surface of the material, especially from the lead. This behavior becomes ever more complex as the thickness of the shield increases. It is readily apparent in these Monte Carlo simulations that the dose is very dependent on the shape (or thickness) of the shield. This scattering of the electrons and their byproducts by the shielding means that the details of geometry of the shielding must be considered in any radiation calculations.

To ease the computational burden, analytical expressions have been fitted to the results of the Monte Carlo calculations and to actual measurements for electrons, protons, heavy ions, neutrons, and photons. Specific characteristics, such as energy deposition, ionization, flux (both forward and backscattered), and dose, then can be predicted as functions of shielding thickness or material. Figure 6.6 illustrates

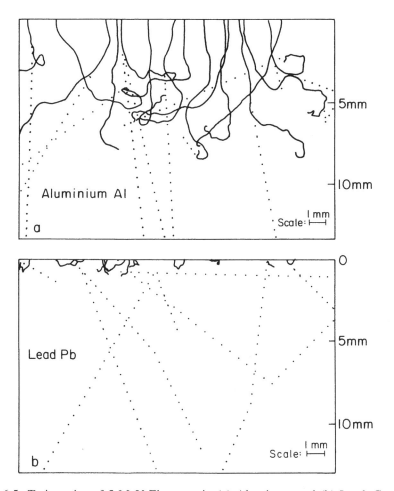

Figure 6.5. Trajectories of 5-MeV Electrons in (a) Aluminum and (b) Lead, Computed Using the GEANT Monte Carlo Code. Electrons are Normally Incident from Above. Dotted Lines Indicate Photons Induced by the Electrons. Used by Permission of Holmes-Siedle and Adams, from Handbook of Radiation Effects, 1993, by Permission of Oxford University Press.

this process for one characteristic, namely the electron dose relative to distance into the shielding material as the incident electron energy is increased. Here, the region over which the electron deposits its energy is smeared out along the track. Contrast this with Figure 6.7 for a proton that deposits its energy primarily near the end of its track. This difference in energy deposition with shielding thickness is often used in designing solid-state particle detectors capable of discriminating between high-energy electrons and protons. It also must be kept in mind when designing shielding that too much shielding could cause cosmic rays of a particular energy to deposit most of that energy at a specific point in a device rather than

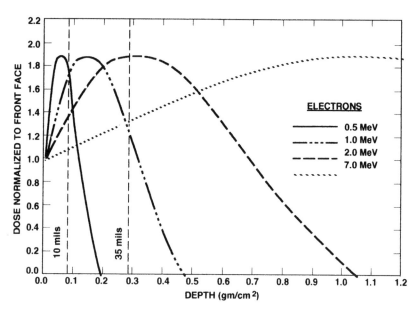

Figure 6.6. Electron Depth–Dose Profiles in CaF$_2$ (with permission of E.G. Stassinopoulos).

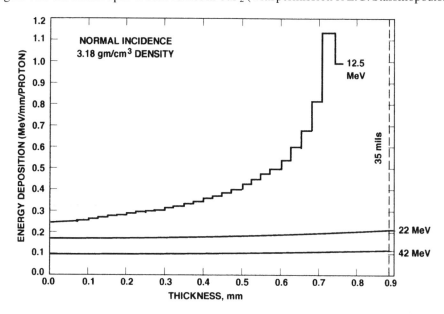

Figure 6.7. Proton Depth–Dose Profiles in CaF$_2$ (with permission of E.G. Stassinopoulos).

passing through it. Families of similar parametric curves for relevant particles and characteristics have been developed that allow rapid calculations of the effects of shielding (lengthy Monte Carlo calculations are, however, often retained in detailed shielding studies).

These considerations allow a simple description of how the radiation-dose environment is determined inside a spacecraft. If the environment interaction is limited to dosage, then the following, very simple one-dimensional model describes the basic mathematical steps involved:

For dose, assume a target of density ρ, area δA, and thickness $\delta \tau = M/(\rho \delta A)$, where M is the target mass:

(1) First determine the attenuation effects of the shielding on the ambient flux spectrum of the particles of interest.

(2) Compute the attenuated flux (number N of particles per unit area δA normal to the surface) relative to energy at the target surface. Call this $f(E)$ at energy E such that:

$$f(E) = \frac{N(E)}{\delta A}. \tag{6.3}$$

(3) Estimate (i.e., from Figure 6.3) the change in energy δE in crossing the target thickness $\delta \tau$ at the appropriate distance in the shield for a particle of initial energy E:

$$\delta E \simeq \delta \tau \left. \frac{dE}{dx} \right|_E. \tag{6.4}$$

(4) The dose per particle of energy E is given by

$$D(E) \simeq \frac{\delta E}{M} = \frac{1}{\rho} \left. \frac{dE}{dx} \right|_E \frac{1}{\delta A}. \tag{6.5}$$

(5) The total dose at energy E is then given by

$$D_T(E) \simeq N(E)D(E) = \frac{1}{\rho} \left. \frac{dE}{dx} \right|_E f(E). \tag{6.6}$$

(6) The total dose for $E > E_o$ is then given by integrating over the range E_o to ∞.

This process, repeated for many different angles and particles, gives the total dose inside a three-dimensional volume. The final answer is basically independent of the shape or size of the test point and is only a function of the density of the material.

In actual dosage calculations, because of the various effects of shielding on the energy deposition, five shielding geometries are typically considered [these definitions are adapted from the NOVICE code (Jordan, 1987)]. These are illustrated in Figure 6.8 and described as follows:

(1) Spherical Shell: As the name implies, this configuration represents a hollow sphere of equal thickness in every direction from the dose site which is at the center of the sphere. (*Note:* The radius of the sphere void can be shown to be unimportant for large distances.) The dosage tends to be lower than for a solid sphere of the same shield thickness. This case resembles a point inside a typical hollow spacecraft.

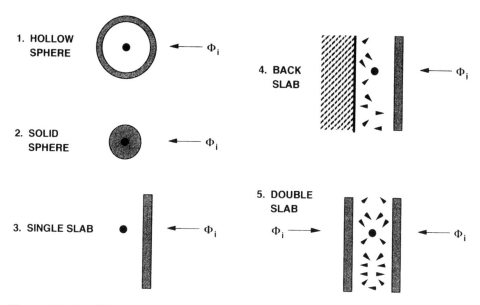

Figure 6.8. Five Shielding Configurations Considered in the NOVICE Code for Calculating Dosage.

(2) Sphere: The shield is assumed to be uniformly distributed around the dose site with no gap between the shielding material and the dose site (i.e., a point at the center of a solid metal sphere). Because scattering takes place relatively close to the dose site, little scattered flux is lost. This case resembles a "spot shield" configuration.

(3) Slab (or 2*Slab): A single slab is assumed to be an infinite two-dimensional surface (Figure 6.8). Ideally, particles enter from one side and irradiate the dose site. That is the basic Single-Slab configuration which assumes no backscattering of electrons and no flux from behind (i.e., infinite back shield). This approximates the actual case for high-energy protons and heavy ions. To estimate the omnidirectional flux for a part between two shield planes without backscattering from a second surface, this value is typically doubled. (In the NOVICE code, this is called the 2*Slab case; see also Double Slab case.)

(4) Back Slab: This configuration is similar to the Slab in that the dose site is again assumed to be backed up by an infinite slab. As before, the flux only comes in from one side but now particles can be reflected or scattered back. This often nearly doubles the incident flux for electrons.

(5) Double Slab: Here, there are two identical thin shields – one on each side. In this configuration, the flux is assumed to come from both sides and backscattering from each is included. This geometry would resemble the case of a flat solar-array panel extending out from the spacecraft in a wing configuration.

Which configuration to use depends greatly on the geometry of the spacecraft component being modeled. The spherical shell is often used as the baseline representation because it more closely resembles the shielding around typical circuit boards in the spacecraft interior.

Consider next the detailed steps involved in determining the other radiation quantity of interest, the Heinrich curve for a heavy ion. As outlined in Adams, Letaw, and Smart (1983) for the CREME code, the steps are as follows:

(1) First define the particle spectrum of interest at the surface of the critical volume. For example, consider the ambient environment for GCR iron at the surface of a spacecraft. Call this differential spectrum $f(E)$.

(2) The attenuation of a high-energy ion by the shielding can be approximated by [spallation is ignored (Adams et al., 1983)]:

$$f'(E') = f(E)\frac{S(E)}{S(E')}\exp(-\sigma\tau), \tag{6.7}$$

where

$$\sigma = \frac{5\times10^{-26}N_A(A^{1/3}+27^{1/3}-0.4)^2}{27}, \tag{6.8}$$

and f' is the differential spectrum inside shielding, τ is the thickness of shielding, E' is the energy inside the spacecraft $\{=R^{-1}[R(E)-\tau]\}$, $R(E)$ is the range through the shield of an ion of energy E (see Figure 6.9 for Fe in Al), R^{-1} is the inverse function of

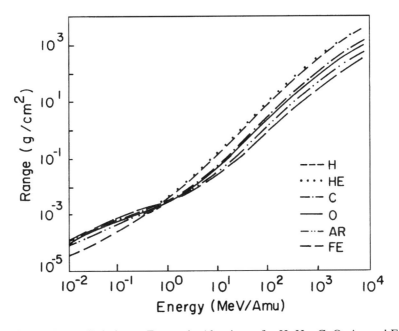

Figure 6.9. Ion Range Relative to Energy in Aluminum for H, He, C, O, Ar, and Fe. The Range is in Units of g/cm^2 and the Energy is in Million Electron Volts per Atomic Mass Units.

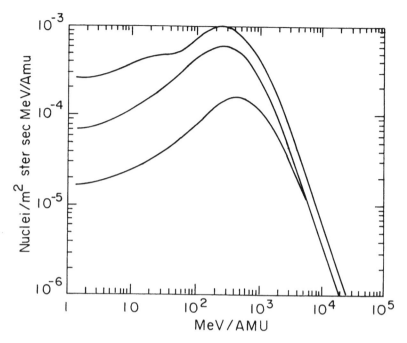

Figure 6.10. Shielding-Attenuated Cosmic Ray Differential Iron Spectra for Three Cases: 90 Percent Worst Case (Upper), Solar Minimum (Middle), and Solar Maximum (Lower). These Spectra are for 1 AU (No Magnetospheric Shielding) and Behind 0.025 Inch of Aluminum Shielding. (Adams et al., 1983).

$R(E)$, $S(E)$ is the stopping power or LET of an ion in the target material (see Figure 6.3 for Fe in Si), A is the atomic mass of the ion, and N_A is Avogadro's number. Figure 6.10 shows the results of this calculation for iron behind 635 μm of aluminum shielding (Adams et al., 1983).

(3) Next, the dE/dx curve for the incident particle species in the material of interest is determined. The $-(1/\rho)dE/dx$ curve for Fe in Si is plotted in Figure 6.3.

(4) The incident (internal) differential particle spectrum $f'(E')$ is converted to the differential Heinrich spectrum $h(\text{LET})$ by

$$h(\text{LET}) = f'(E')\frac{dE'}{d\text{LET}}.$$ (6.9)

(*Note:* A given LET may correspond to several values of E'.)

(5) Equation (6.9) is integrated over LET to give the integral Heinrich LET curve,

$$F_H(\text{LET}) = \int_{\text{LET}}^{\infty} h(\text{LET})d\text{LET}.$$ (6.10)

This equation is equivalent to Equation (2.89). The final results are plotted in Figure 6.11. These curves, called Heinrich curves, are normally required for SEU

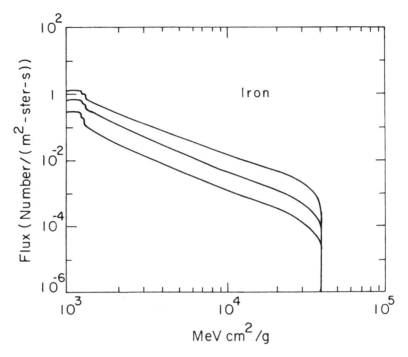

Figure 6.11. The Same Three Cases as in Figure 6.10 are Used, but Now the Spectra have been Integrated to give the Integral LET Spectra for Iron. (Adams et al., 1983).

calculations. In the case of GCR, this curve would be calculated for each GCR species and summed over all of the species to give a composite spectrum.

To summarize, there are many different techniques for estimating the radiation environment behind a spacecraft shield. To limit the amount of computer time required, the "exact" Monte Carlo formalism is often replaced by analytical approximations (called kernels) when performing the particle transport and shielding calculations. Specifically, tabulated attenuation data, using Monte Carlo techniques, are prepared for various shield geometries (i.e., the slab, spherical shell, and solid-sphere geometries illustrated in Figure 6.8). Given a three-dimensional model of the shielding mass and geometry (or a one-dimensional configuration, depending on the desired level of accuracy desired), the equivalent shielding at a point as a function of angle and path length is calculated. The input spectra from the environment (e.g., neutrons, gamma-rays, photons, electrons, positrons, protons, heavy ions, alphas, GCR) are convolved with this equivalent shielding to calculate the dose as a function of energy (or the Heinrich flux as a function of LET) and angle. Secondary and bremsstrahlung particle effects also normally need to be included, particularly for thick shielding. For a more detailed treatment of these interactions, see Holmes-Siedle and Adams (1993) and others.

6.3.1 Solar-Array Degradation

Besides nuclear power, the major long-term means to power a spacecraft is by solar arrays. A solar array is made up of solar cells, each of which is a semiconductor sandwiched between a substrate and a coverglass. Radiation damage can occur to a cell from solar-flare protons, trapped electrons, and trapped protons. As discussed, the damage mainly consists of ionization and atomic displacement. These affect both the voltage and current output of the cell and result in a power degradation in the output of the cell (Agrawal, 1986). This loss of power is a primary cause of satellite failure and thus a major concern for spacecraft design.

For the purposes of comparison for different cell types and radiation particles, it is conventional to express the radiation fluence in terms of the "1-MeV equivalent fluence" (typically, 1-MeV electrons or 1-MeV neutrons in silicon). This fluence is defined by

$$\phi_e = \int_{E_{ce}} K_e(E)\frac{d\phi_e(E)}{dE}\,dE + 3{,}000\int_{E_{cp}} K_p(E)\frac{d\phi_p(E)}{dE}\,dE, \tag{6.11}$$

where E_c is the cutoff energy for the coverglass (i.e., the minimum energy of the electron or proton necessary to penetrate the coverglass), K_e is the damage coefficient for 1-MeV electrons, K_p is the damage coefficient for 10-MeV protons, and $d\phi_{e,p}/dE$ is the electron, proton differential fluence. The damage coefficient is defined as the factor relating the equivalent fluence of 1-MeV electrons (or 10-MeV protons) to electrons (protons) at energy E necessary to give the same level of damage. For a flat sun-oriented solar array at GEO, the average equivalent 1-MeV fluence after seven years is 10^{15} electrons/cm^2, whereas for a spin-stabilized satellite that has body-mounted arrays and therefore has more protection, the average equivalent 1-MeV fluence after seven years is 3×10^{14} electrons/cm^2. In Figure 6.12, the power degradation is shown for three types of silicon cells. Typical power degradation in the output of an array over a seven-year lifetime is 15 to 20 percent. Satellite system engineers must make a complex trade-off in designing a GEO satellite between a certain amount of power degradation and increasing the shield (coverglass) thickness. Increasing the thickness increases the mass of the array, and for a fixed-mass satellite decreases the mass budget for the payload.

6.3.2 SEUs Due to Heavy Ions

Single Event Effects (SEEs) result when the passage of a single, high-energy particle introduces unwanted ionization in a circuit element or sensor. These effects can range from a simple light flash in an optical sensor to the creation of a sneak circuit in an integrated circuit element, resulting in the shorting and burnout of the

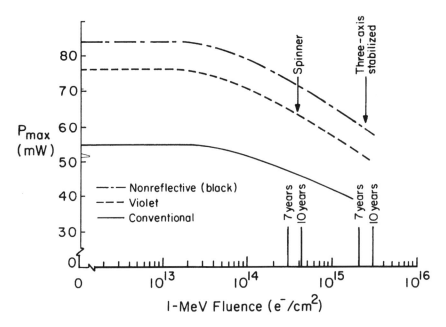

Figure 6.12. Maximum Power Output of Conventional, Violet, and Nonreflective Cells as a Function of 1-MeV Electron Fluence for a 2 × 2 cm Cell.

element. Here, for simplicity, only the most common form of SEE, the so-called single-event upset or SEU is considered. An SEU is produced when a high-energy charged particle passes through an element (or elements) of an integrated circuit and produces a change (or changes) in a memory element or in the logic state. The basic process and computations associated with SEUs are generally applicable (with minor changes) to the other classes of SEEs and serve as a useful introduction to SEEs.

At the most basic level, an SEU occurs when a high-energy particle (typically heavy ions as electrons and protons do not normally deposit sufficient ionization) passes through the "critical volume" (also called the depletion region) of an IC memory element. This ionization creates electron-hole pairs along the particle track. These charges become separated by the electric field across the critical volume or depletion region and are attracted to the nodes of the element. When they reach a node they cause a short current pulse. If the total charge in the resulting pulse is larger than a characteristic critical charge (proportional to the electric charge stored in the memory element) and the pulse is of the right duration, it will cause a change in the state of the element. Figure 6.13 (Robinson, 1988) illustrates this passage of a heavy ion through the critical volume of a typical memory element. Funneling of the charge (a process by which extra charge outside the original ionization channel is pulled into the channel) and charge diffusion (the subsequent slow migration of

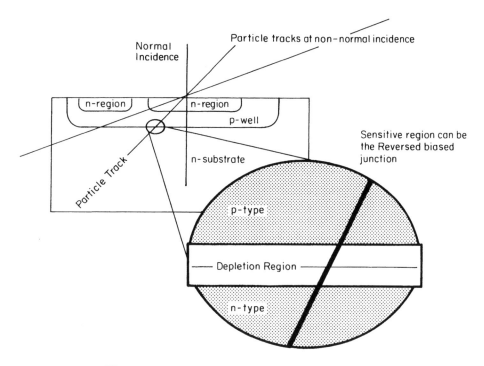

Figure 6.13. SEU Upset Diagram. (Robinson, 1988).

any remaining charge along the particle path) will increase the charge collected over the initial pulse.

SEUs primarily occur in components with some form of memory retention. Memory devices are generally classed as charge storage devices, voltage storage devices, or current steering devices. Devices that use stored charge (e.g., dynamic RAMs and charge-coupled devices) determine their memory state by the presence or absence of that charge. Voltage storage devices (e.g., static RAMs or CMOS RAMs) determine their state by the voltage present at each node in a flip-flop circuit. Bipolar devices establish their memory state by steering currents such that certain transistors are in an "on" state. Bipolar technologies for memory and processor applications include transistor-transistor logic (TTL) and integrated injection logic (I^2L). All are susceptible to SEUs (Robinson, 1988).

Consider first the issue of critical charge. Prior to the 1970s, SEUs were not a problem. They have become a problem, however, as the drive toward lower power and higher speed ICs has lead to smaller and smaller levels of critical charge. ICs store information using a total charge in the range of 0.01 picocoulomb to 1.0 picocoulomb. Feature sizes such as the depletion depth (the length over which the electric field extends) are at present on the order of 1 μm. A high-energy ion with

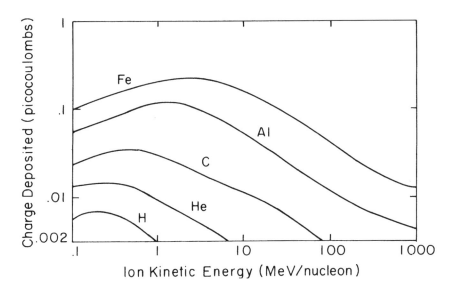

Figure 6.14. Charge Deposited in 1 μm. (Robinson, 1988).

an energy between 0.1 and 10^3 MeV/nucleon (typical GCR energies) will deposit a charge of this order in 1 μm of silicon (see Figure 6.14 from Robinson, 1988). As future devices will have critical charges down to 10^{-3} picocoulombs, the SEU problem is likely to grow with time. Already, the critical charge of a typical circuit element is at the point where the passage of a GCR or other high-energy heavy ions can deposit sufficient charge to generate a bit flip. Indeed, SEU-induced bit flips and the resulting changes in operational state are now suspected of causing false commands such as power-on resets in several spacecraft.

Although the response of a circuit is difficult to compute and dependent on the actual circuit layout, the process can be readily summarized. As discussed earlier, the current injected is the sum of the prompt charge (the original charge directly deposited in the critical volume or depletion region), the charge swept up in the so-called funnel, and the charge that diffuses slowly from the ion path into the depletion region. The prompt and funneled charges are separated and collected within a nanosecond or less. The delayed diffusion component is collected over one to hundreds of nanoseconds. Thus the charge pulse injected by the ion appears as a short pulse with a nanosecond or less decay, followed by a long-duration, low-amplitude current coming from the charge diffusing along the original ion track. The normal, circuit-induced pulse necessary to change the state of the element also has a sharp rise and fast fall time with most of the charge collected in less than a nanosecond. Therefore, as long as the ion-induced pulse is shorter than the normal circuit response time, the critical charge is basically independent of pulse shape.

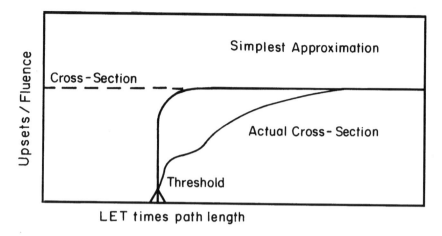

Figure 6.15. Classic-Experiment Cross Section. (Robinson, 1988).

Quantitatively, the second problem, that of determining the SEU rate, reduces to determining the size (and shape) of the critical volume, determining the critical charge deposited by each particle, and then convolving these with the flux of high energy particles. The sensitive volume is determined by the sensitive element's feature size, normally expressed in terms of a cross-section and thickness or depth (typically 1 to 10 μm). Figure 6.15 (from Robinson, 1988) is a schematic of the cross-section as function of the particle's LET. To first order, the SEU cross-section can be assumed to be a step function with a threshold determined by the charge stored in the element (unfortunately, the actual cross-section, as shown, can vary with LET, complexly shaped sensitive volumes, funneling, impact angle, ion species, and the operational state of the element).

To generate an SEU, the ionizing particle must deposit a charge equivalent to the critical charge. In addition to the shape of the region (which determines the length over which the particle can deposit charge), the charge density per unit path length must be determined. The critical quantity here is the energy required to produce an electron-hole pair in the material making up the element (usually silicon or gallium arsenide). For silicon, this is 3.6 eV/electron-hole pair (gallium arsenide is 4.8 eV/electron-hole pair). That is, for every 3.6 eV of energy deposited by the passage of the particle, one electron and one hole (or 1.6×10^{-19} coulombs per unit charge) are produced. The total charge deposited is then determined by multiplying the path length of the particle through the critical volume by the particle's dE/dx or LET (also called stopping power) times the density of the material. If this charge exceeds the critical charge, then an SEU will occur.

Finally, the total upset rate is calculated by summing, for all angles, the number of particles passing through the critical volume that can cause an SEU. This is often done by using Monte Carlo techniques that take the three-dimensional shape of the critical volume and the shielding of the structure into account. Here, a simpler formulation will be discussed that evaluates the integral of the product of the cross-section (which contains the information on the path-length distributions through the critical volume) and the ion flux at the element as functions of LET (rather than energy), spatial parameters, and ion species. The integral representation of the SEU rate, dU/dt, can then be simplified as follows:

$$\frac{dU}{dt} = \int_0^{2\pi} d\phi \int_0^{\pi} \sin\theta d\theta \sigma(\text{LET}, \theta, \phi) \sum_{Z=1}^{Z=92} F_H(\text{LET}), \qquad (6.12)$$

where $F_H(\text{LET})$ is the Heinrich flux as a function of LET as defined in Eq. (6.10) and σ is the cross-section as a function of LET and angle (θ, ϕ) (all variations with species have been ignored). The summation is over all relevant ion species. The most difficult part of this computation is determining the function σ so that it properly accounts for the variations with angle and takes into account the probabilities of all path lengths. For simple geometries (i.e., thin parallelepipeds), normal incidence is usually all that is considered and σ is then a one-dimensional function of LET, determined by experiment, as illustrated in Figure 6.15. There are, however, path lengths (from one corner on the front diagonally to an opposite corner on the back) that could be very long and result in enhanced deposition. Fortunately the cross-section is typically small for high angles of incidence (this might not be true for complex cross-sections).

As has been discussed, the principal environment that produces SEUs is the high-energy, heavy ion component of the solar proton events, the trapped radiation belts, and the GCR. Heavy ions are indeed present in most radiation environments. Although shielded by planetary magnetic fields, GCR ions will be minor for almost all missions. Solar proton events, even though they produce heavy ions at energies usually lower than GCRs, are usually much more of a concern because of their often high fluences. The trapped radiation heavy ion environment is dependent on the particular planet with the Earth and Jupiter being the worst examples. The natural SEU-causing environments are summarized in Table 6.1 (from Robinson, 1988). The anomalous component refers to singly ionized ions which are occasionally seen as a function of solar cycle and increase the background flux for those ions in the 10- to 100 MeV/amu range.

A final and likely critical issue for future space components will be their sensitivity to protons. At present, space-qualified parts normally have a critical LET of $\sim$10 MeV-cm^2/mg or higher. This is too high in general for ambient protons in the Earth's environment to cause SEUs directly as they have fluxes peaking at LETs well below this level. Therefore, until recently, most parts were considered to be relatively immune to SEUs by direct proton ionization unless the impact was at

Table 6.1. *Natural SEU-causing environments (Robinson, 1988)*

Natural Environments	Time-dependent	Position-dependent	Comments
Galactic cosmic rays	yes	no	~1 GeV/amu $Z \geq 1$
Anomalous component	yes	no	~20 MeV/amu He, N, O, Ne
Planetary radiation belts	yes	yes	protons; ~50 MeV; some heavy ions
Solar particle events	yes	yes	protons; ~100 MeV; some heavy ions

a glancing angle (i.e., a long path length as discussed earlier) – a low probability event. This is fortuitous as the proton fluence near the Earth is 10^4 to 10^7 higher than the fluences of heavier ions. Protons may, however, cause SEUs indirectly by nuclear reactions or knock-on that generate energetic heavy ions within the sensitive volume. These secondary ions can then cause the SEU. The likelihood of this type of interaction is determined by the Bragg peak for the material. The Bragg peak for silicon is 17 MeV-cm^2/mg implying that the proton-secondary ion SEUs will not be significant unless the cross-section threshold for the device is below about 8 MeV-cm^2/mg (Robinson, 1988). As parts' critical LETs fall below ~8 MeV-cm^2/mg in coming years, expect to see increasing problems with SEUs in space.

6.4 Radiation Charging of Dielectric Materials

Spacecraft charging as discussed in this book is typically assumed to mean the buildup of charge on the exterior spacecraft surfaces. In reality, except for bulk-conductive surface materials, charge will be deposited over a finite depth – indeed, any particle with energy over a few electron volts will penetrate the surface. The depth of penetration and charge deposition are functions of stopping power, the energy of the impinging particles, and any electric fields normal to the surface. A common spacecraft surface configuration that will exhibit this behavior consists of an exposed dielectric material with a conducting backing connected to the spacecraft ground (e.g., optical coatings, solar-cell coverglasses, or exposed insulated wiring). Charge will accumulate (or diffuse away) in the dielectric over time as a function of the conductivity of the material and the imposed electric fields. If the charge accumulating in the dielectric induces a field greater than the breakdown strength of the material, which typically is of the order 10^5 to 10^6 V/cm, then a discharge can occur in the material (as opposed to between external surfaces) or from the interior of the dielectric to one of its surfaces. Because this discharge is potentially damaging to the spacecraft – its surfaces and its systems – it is of interest to the spacecraft designer.

The bulk discharge process often has been observed for dielectrics exposed to high-energy radiation environments on the ground. High-energy electrons, for example, produce a characteristic "Lichtenberg" discharge pattern (in clear plastics, this pattern looks like a frozen lightning strike if the discharge reaches the surface and emits material). In addition to this material "crazing," the discharge produces effects on nearby electronic systems just as in the case of a surface discharge. Because the discharge can occur interior to the exterior EMI shielding or directly to supposedly shielded wires, it often can do much more damage than an exterior discharge. Discharges may release material – typically gases and plasma – that can, for certain configurations, trigger surface arcs (Fredrickson, 1980). Thus radiation charging of dielectrics or bulk charging (often called buried or internal charging) may be a major source of arc discharges. Evidence from *SCATHA* and *CRRES* indeed indicate that as much as 50 percent of all arc discharges are probably caused by this phenomenon as opposed to surface charging (Vampola, 1987). Given the importance of the phenomenon, its characteristics are reviewed in Section 6.4.1, with emphasis on its physics.

6.4.1 Physics of Radiation-Induced Charging

The computations involved in estimating radiation-induced charging resemble surface-charging calculations with the inclusion of space charge. That is, the basic problem is the calculation of the electric field and charge density in a self-consistent fashion over the three-dimensional space of interest. The primary difference is the role that the conductivity of the material plays in the process. As before, Poisson's equation must be solved subject to the continuity equation, but now in the dielectric. As a simple example, consider a one-dimensional, planar approximation at a position x in the dielectric. The equation at x is then

$$\epsilon \frac{dE}{dt} + \sigma E = J, \tag{6.13}$$

where E is the electric field at x, t is time, σ is the conductivity in ohm^{-1} m^{-1} = σ_o + σ_r. Here, σ_o is the dark conductivity (conductivity in the absence of illumination by radiation), σ_r is the radiation-induced conductivity (Fredrickson et al., 1986), ϵ is the dielectric constant of the material given by $\epsilon = \epsilon_{rel}\epsilon_0$, where ϵ_0 is the free-space permittivity = 8.8542×10^{-12} F-m^{-1}, and ϵ_{rel} is the relative dielectric constant. Finally, J is the incident particle flux (current density) at x including primary and secondary particles.

This follows from Poisson's equation and current continuity with the total current consisting of the incident current J (primary and secondary particles) and a conduction current σE. A solution of this equation for σ and J independent of time is

$$E = E_0 \exp(-\sigma t/\epsilon) + (J/\sigma)[1 - \exp(-\sigma t/\epsilon)], \qquad (6.14)$$

where E_0 is the imposed electric field at $t = 0$.

This is only a crude approximation to reality because geometric effects, time variations in the conductivity and incident current, and other effects make numerical solution a necessity. It is, however, useful in understanding the time constants ($\tau = \epsilon/\sigma$) involved in charging the dielectric. Typical values for τ range from 10 s to 10^3 s for $10^{-16} < \sigma_o < 10^{-14}$ (ohm-m)$^{-1}$. In regions where the dose rate is high (enhancing the radiation conductivity), the E field comes to equilibrium rapidly. In lightly irradiated regions, where the time constant is long (the dark conductivity dominates), the field takes a long time to reach equilibrium.

The peak electric field (E_{max}) in an irradiated dielectric is estimated [e.g., Fredrickson (1980)] for radiation with a broad energy distribution to be

$$E_{max} = (A/k)/(1 + \sigma/kD) \approx (A/k), \qquad (6.15)$$

where $A = 10^{-8}$ s-V/ohm-rad-m^2, $k =$ coefficient of radiation-induced conductivity in s/m-ohm-rad, and $D =$ average dose rate in rad/s.

The second approximation follows for high flux conditions because σ_r can be approximated by $\sigma_r \approx kD^\delta$ [where $\sigma_r > \sigma_o$ for high fluxes and $\delta \approx 1$, Fredrickson (1980)]. The equation is in agreement with analytical solutions when they exist and, for some configurations, more complex numerical solutions. Typical values of k are $10^{-16} < k < 10^{-14}$ for polymers [Fredrickson (1980)]. Inserting the range of values for k, E_{max} can reach up to 10^6 to 10^8 V/m – the range where breakdowns are expected.

Although particles with energy as low as 5 eV can deposit charge, electrons with energies of 100 keV or higher are normally assumed to be the primary sources of buried charge. First, protons and heavier ions require energies in excess of 100 keV to penetrate more than a few micrometers and, at these energies, their fluence typically is too low to deposit significant charge in a short time. On the other hand, 100-keV electrons penetrate ≈ 70 μm. Further, electrons between 100 keV and several million electron volts can both penetrate deeply into the dielectric and have enough fluence over a short time to build up sufficient charge to cause a breakdown. This critical fluence can be approximated by assuming that all of the charge is trapped on an ideal parallel-plate capacitor having an imposed electric field, E, of 10^7 to 10^9 V/m. If the dielectric constant for the capacitor is ϵ, then

$$Q = CV = \epsilon AV/d = \epsilon AE, \qquad (6.16)$$

where Q is the total charge, C is the capacitance, V is the potential between plates $= E/d$, A is the plate area, d is the distance, $\epsilon = \kappa\epsilon_o$, and $\kappa \approx 1\text{--}10$.

For 1 coulomb $= 6.25 \times 10^{18}$ electrons, the above implies that the critical fluence must be $Q/A \approx 5 \times 10^{10}$ to 5×10^{11} electrons/cm^2. For a flux of 5×10^6 electrons/cm^2 s [the maximum flux at 1 MeV for geosynchronous orbit, see Vampola (1987)], the time period would be 10^4 to 10^5 s (or 3–30 hours). Measurements of the $E > 1.2$-MeV electrons at geosynchronous orbit by the the GEOS-2 satellite between July 1980 and May 1982 and star-sensor anomalies on the U.S. Air Force (USAF) Defense Support Program satellite have been found to be correlated. The strong correlation between these anomalies and the high-energy electrons is interpreted as evidence of this type of discharge phenomenon.

6.4.2 Experimental Evidence for Radiation-Induced Bulk Discharges

To test these concepts and verify the process of dielectric or radiation-induced charging, the internal discharge monitor, or IDM (Fredrickson, Holeman, and Mullen, 1992) was flown on the *CRRES* satellite to monitor radiation-induced pulses. A variety of configurations and materials were flown. Over 4,000 pulses were detected during the 13-month lifetime of the *CRRES* mission (Fredrickson et al., 1992). A few of the main results of the IDM experiment, as summarized by Fredrickson et al. (1992), are presented in the following:

(1) Pulsing of dielectrics is common in the electron belts when the flux exceeds a fixed value (5 nA/m^2). The rate was weakly proportional to the electron flux above this level.
(2) Protons did not appear to correlate with the pulses (their effects may have been swamped by the electron effects).
(3) Thicker samples and those with more exposed insulator pulsed more often.
(4) The pulsing characteristics changed with time (exposure). After seven months. the pulse rate averaged almost six times more than during the first seven months.
(5) The pattern of pulse changes over time varied with material. The TFE Teflon samples decreased their pulse production after exposure to 10^6 rads. The other samples, including FEP Teflon, showed enhanced pulsing after this exposure.

The IDM thus demonstrated the effects of buried charge and long-term effects of radiation on the material properties (charge buildup and discharging).

In late January 1994, two geostationary communications satellites, Anik E-1 and E-2, suffered anomalies in their attitude control circuitry during an enhanced high-energy electron event at GEO. Ground controllers were eventually able to regain control of the Anik E-1 but lost Anik E-2 for several months. Although not conclusive (the same environmental conditions have been encountered several times in the past without problems), there is strong evidence that the cause was internal or radiation-induced charging. At a minimum. the event cost millions in lost communications and personnel costs.

The steps taken to limit the effects of radiation-induced charging on spacecraft are similar to the steps taken for surface charging. In addition, the conductivity of the materials should be enhanced and the size of the source regions limited as much as possible:

- Use conductive materials wherever possible. For dielectric surfaces this translates to enhancing the conductivity of the material (typical values are $\sigma \geq 10^{-10}$ ohm^{-1} m^{-1}).
- Break up potential source regions. For dielectrics such as optical solar reflectors, the guidelines call for dividing the volume/surface area into regions smaller than ≈ 1 cm^2.
- Ground all floating conductors.

When these steps are combined with the precautions taken for surface charging, radiation-induced charging can be reduced as a threat in most cases.

6.5 Radiation Environment Estimates

In this section, the analysis of the spacecraft radiation environment is completed by combining knowledge of the ambient radiation environment and its effects with the transport/shielding process to examine several practical case studies. These are a detailed analysis of the radiation-dose environment to be expected for a lunar transfer mission and a comparison of the intense radiation environment at Jupiter with the saturated nuclear environment at the Earth. The case studies illustrate the basic steps required in carrying out a thorough analysis of the radiation environment inside a spacecraft. Heinrich flux estimates for several environments and a discussion of recent results from the *CRRES* program conclude the section.

6.5.1 Example: The Clementine Program

As an illustration of the process of estimating the radiation-dose environment within a satellite, consider the case of the *Clementine* spacecraft lunar transfer orbit sequence. *Clementine* was an ambitious Department of Defense (DoD)/NASA mission designed to map the Moon. It was launched in January 1994 and was the first mission by the United States to the Moon in over 20 years. *Clementine* tested the effects of the radiation environment on a number of unique, advanced microelectronic systems. *Clementine* also left behind its lunar transfer stage in an unusual, highly elliptical orbit. This *interstage* and the *Clementine* were both instrumented with radiation dosage and SEU detectors. In addition, each carried boxes of advanced microelectronics components for direct exposure to the radiation environment. A detailed radiation environment prediction was required to allow the identification of radiation-sensitive parts and to determine appropriate replacement parts or provide enhanced protective measures. It was also desired to predict the performance of the systems and to test components in the radiation environment. These are all typical requirements for a space mission and illustrate the wide range of potential radiation applications.

6.5.1.1 AE8 and AP8 Radiation Dosage Results

The first step in the determination of the *Clementine* mission dosage was the computation of the radiation dosage due to the Earth's trapped radiation environment. As discussed in Chapter 3, the most often used trapped radiation environment models are the AE8 (electron) and AP8 (proton) solar maximum (or active) and solar minimum (or quiet) trapped radiation models. These models give dosage results that, when averaged over mission lifetimes of the order of the solar cycle, are typically within a factor of two of the actual measured dosages. Unfortunately, for time periods shorter than about five years, the statistical variations can be great (approaching factors of 10 to 100 for missions of less than a year, such as *Clementine*). Even so, they have formed the basis of almost all trapped-radiation calculations since the late sixties. With a properly defined radiation design margin (RDM), their predictions are useful in evaluating a spacecraft radiation-hardness design.

As has been described, the process of calculating the dosage at the interior of a spacecraft is straightforward but time-consuming. First, the B and L coordinates of the spacecraft are estimated from the orbit. The particle integral flux as a function of energy is then computed in terms of B and L from the AE8 and AP8 models. The resulting spectra are summed over mission time to give the total integral fluence spectrum in terms of energy and particle species. These spectra, by species (electrons and protons), are then used as input to the shielding code (the NOVICE code, a commercially available software package, was used here), which computes the total energy deposited at a point as a function of shield thickness, shield composition, and geometry. Typically, for dosage calculations to be used in a first-order estimate of the internal radiation environment, aluminum for the shield and silicon for the dose site are assumed for composition. Geometric considerations become particularly important for electrons because the electrons can be easily scattered or reflected within the material. As described in earlier sections, several different geometries are usually assumed. Here, five geometries are considered (see Figure 6.8): (1) Spherical Shell, (2) Sphere, (3) 2*Slab, (4) Back Slab, and (5) Dubl Slab (Double Slab).

In the NOVICE calculations, the single slab is assumed to be an infinite two-dimensional surface (Figure 6.8) with an infinite back shield, so that no radiation comes from behind the shield and none is reflected back. For comparison purposes, the code doubles this value (hence, 2*Slab) so that the results can be used to estimate the omnidirectional flux for a part between two slabs without scattering (see Dubl Slab case). Unless stated otherwise, the spherical shell geometry is assumed to be the baseline representation because it more closely resembles the shielding around typical circuit boards in the *Clementine* interior.

Table 6.2. Clementine *orbit phase*

Orbit phase	Date
1. Low Earth orbit	94/01/24
2. Earth–Moon trans	94/01/26
3. Earth–Moon orbit	94/02/04
4. Earth–Moon orbit	94/02/15
5. Earth grav assist	94/05/06
6. Earth grav assist	94/05/24

A set of orbital data for the portions of the *Clementine* orbit within 11 Earth radii was assembled. The *Clementine* orbit was then divided into six segments. These are given in Table 6.2. For these orbits, the dosage for the five different geometric configurations, solar maximum and solar minimum environmental assumptions, and various shielding thicknesses were calculated so as to provide an estimate of the range of doses to be expected. Radiation doses for electrons, protons, and photons (secondary particles) were calculated for all segments when *Clementine* would be inside $11R_E$ (the AE8 models are only useful inside $11R_E$ and the AP8 models within $6R_E$). The results in terms of dosage are presented in Figures 6.16

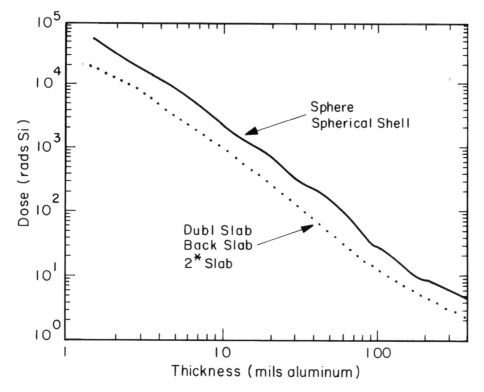

Figure 6.16. Radiation Dosage from the Trapped-Proton Environment (AP8) for Solar-Quiet Conditions and for the First Earth–Moon Transfer Orbit.

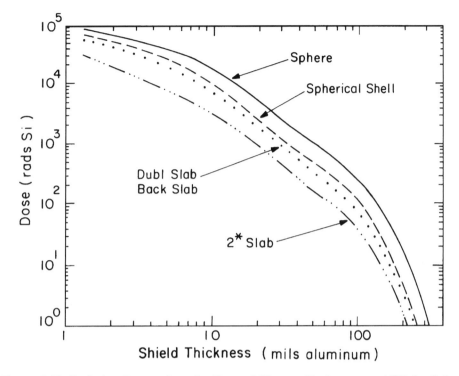

Figure 6.17. Radiation Dosage from the Trapped-Electron Environment (AE8) for Solar-Active Conditions and for the First Earth–Moon Transfer Orbit.

and 6.17 for the worst-case orbit segments: the trapped electrons at solar maximum and protons at solar minimum along the transfer orbit. Dosages are tabulated in Table 6.3 for 58 mils of aluminum shielding for all the orbital segments in Table 6.2.

Note the wide variation in results for the electrons. As has been discussed, the precise geometric assumptions can make a large difference in the estimates. The main difference between the proton results is that between the two-dimensional and three-dimensional geometric assumptions of a plane versus a sphere. In many cases, the spherical-shell geometry is the most appropriate because it resembles the structure of a spacecraft surrounding circuit boards in the interior of the spacecraft. Spherical-shell results for the *Clementine interstage* mission are summarized in Figure 6.18. (The *interstage*, left behind in the more stressing lunar transfer orbit, was to have had a perigee of 500 km, an apogee of 160,000 km, and an inclination of 67 degrees. In the actual case, the apogee was closer to 130,000 km). Note that the proton dose due to trapped radiation is very low in comparison to the trapped electrons for the *interstage*.

The solar-flare proton environment was potentially the most severe radiation environment for a lunar transfer orbit such as that of *Clementine* and its *interstage*. The reason is that the spacecraft typically spend little time in the trapped-radiation

Table 6.3. *Summary table for dosage behind 58-mil shielding for the trapped-electron (AE8) and trapped-proton (AP8) environments for* Clementine.

P	SPHERE	SPH SHELL	2*SLAB	DUBL SLAB	BACK SLAB
			Geometry		

Active electrons, dosage (rads)

P	SPHERE	SPH SHELL	2*SLAB	DUBL SLAB	BACK SLAB
1	0.001150	0.000437	0.000154	0.000315	0.000315
2	184.000	83.500	32.600	61.300	61.400
3	778.000	359.000	127.000	260.000	260.000
4	466.000	201.000	78.400	143.000	143.000
5	247.000	88.400	30.100	60.400	60.400
6	665.000	267.000	107.000	183.000	183.000
$\sum$	2340.001	998.900	375.100	707.700	707.800

Active protons, dosage (rads)

P	SPHERE	SPH SHELL	2*SLAB	DUBL SLAB	BACK SLAB
1	0.001800	0.001800	0.001310	0.001310	0.001310
2	29.000	29.000	9.780	9.780	9.780
3	104.000	104.000	31.900	31.900	31.900
4	66.800	66.800	19.400	19.400	19.400
5	0.000	0.000	0.000	0.000	0.000
6	0.000	0.000	0.000	0.000	0.000
$\sum$	199.802	199.802	61.081	61.081	61.081

Quiet electrons, dosage (rads)

P	SPHERE	SPH SHELL	2*SLAB	DUBL SLAB	BACK SLAB
1	0.001120	0.000434	0.000153	0.000312	0.000312
2	128.000	56.300	21.700	41.400	41.400
3	557.000	250.000	87.500	181.000	181.000
4	466.000	201.000	78.400	143.000	143.000
5	247.000	88.400	30.100	60.400	60.400
6	665.000	268.000	107.000	183.000	184.000
$\sum$	2063.001	863.700	324.700	608.800	609.800

Quiet protons, dosage (rads)

P	SPHERE	SPH SHELL	2*SLAB	DUBL SLAB	BACK SLAB
1	0.001800	0.001800	0.001310	0.001310	0.001310
2	29.100	29.100	9.830	9.830	9.830
3	104.000	104.000	32.000	32.000	32.000
4	66.700	66.700	19.400	19.400	19.400
5	0.000	0.000	0.000	0.000	0.000
6	0.000	0.000	0.000	0.000	0.000
$\sum$	199.802	199.802	61.231	61.231	61.231

	SPHERE	SPH SHELL	2*SLAB	DUBL SLAB	BACK SLAB
$\sum_{active}$	2539.80	1198.70	436.18	768.78	768.88
$\sum_{quiet}$	2262.80	1063.50	385.93	670.03	671.03

Note: Results correspond to differing levels of geomagnetic activity, geometry, and orbit. Units are rads (Si). The orbit phase (P) is described in Table 6.2.

Table 6.4. *Worst-case solar proton dosage predictions for one-year*
Clementine *mission scenario*

Probability (%) that dose will not be exceeded	Maximum dose [rad (Si)]
50	466
90	2,750
95	4,640
99	12,700

Note: A 58-mil aluminum shell is assumed.

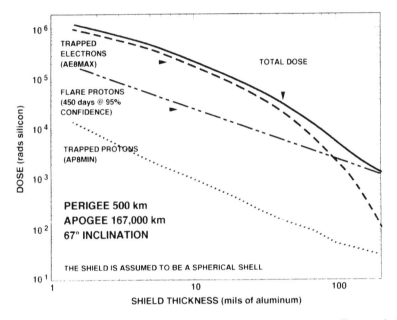

Figure 6.18. Mission Summary Plot for the Total Radiation Dose Expected for the Clementine Interstage. This Assumes a 450-Day Mission and a 95 Percent Confidence Flare Environment. Dosage is Plotted as a Function of Aluminum Shield Thickness for a Spherical-Shell Geometric Configuration.

environment near perigee, whereas more time is spent at apogee far from the Earth and more directly exposed to the solar-flare environment. The dosage expected, as predicted by the JPL model (see Section 3.4.1.3) for a 99% confidence level at 1 AU for solar-flare protons, is presented in Figure 6.19. The results for *Clementine* are summarized in Table 6.4 for a one-year mission within the current active solar cycle (the main *Clementine* mission was from February 1994 to August 1994, while the *interstage* reentered after 100 days). For reference, the next maximum is currently assumed to start on May 26, 1999, and to last seven years. The JPL

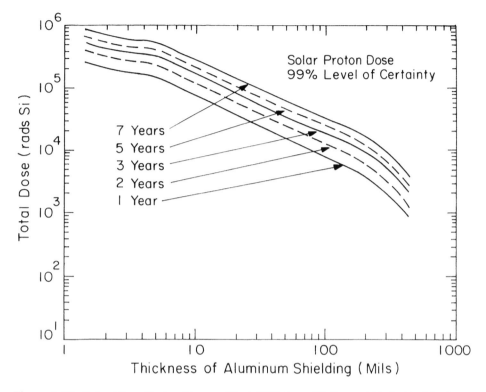

Figure 6.19. Solar Flare Proton Dosage [Rad (Si)] for a 99 Percent Probable Maximum Dose After 1, 2, 3, 5, and 7 Years into the Active Portion of the Solar Cycle Beginning on May 26, 1999. Results are for 1 AU (No Magnetospheric Shielding).

model predicts the likelihood (95 percent for *Clementine*) that a specified maximum integral fluence will not be exceeded in a given time interval (or, vice versa, for a specified likelihood percentage, the corresponding maximum fluence expected). *Clementine*, flying as it did four years after solar maximum, caught the last year of the solar-maximum period. The maximum dosages expected are given by Table 6.4 (assuming one year and 58 mils of shielding).

This section has reviewed the practical steps required to estimate the radiation environment for *Clementine* and its *interstage*. For this environment, each of the major contributors to radiation dosage (i.e., trapped electrons, trapped protons, and solar-flare protons) were identified. From Table 6.3, estimates of the total dose (assuming a 58-mil aluminum shield and a spherical shell) for the trapped environment range from 900 to 2,000 rad (Si) for the electrons and 200 rad (Si) for the protons. From Table 6.4, the solar-flare proton dosage is, for a 95 percent probability that the value won't be exceeded, 4,600 rad (Si). The corresponding dosages for the *interstage* give a total dosage of 10,000 rad (Si) for the electrons, 100 rad (Si) for the protons, and 4,600 rad (Si) for the flare protons (95 percent). Whatever levels are adopted, it is always necessary to establish a project-defined radiation design margin (RDM). For an RDM of four times the predicted dose, this gives 24 krad (Si) as the maximum design dose for a 58-mil aluminum shell for the *Clementine* and 60 krad (Si) for the *interstage*. As a final issue, shielding geometry is clearly a major driver in determining dose, particularly for the electrons. For particularly sensitive components, it is recommended that detailed dosage calculations always be carried out if the initial estimates are a cause for concern. Spot shielding and clever placement of the instruments can significantly lower the dosage to these components without necessitating their replacement.

6.5.2 Example: Jovian Model Application

A Jovian trapped radiation model (Divine and Garrett, 1983) has been developed for several practical applications. These applications, as would be expected, range from establishing radiation dosage guidelines to SEU modeling. Two unique applications are discussed here. The first is the effect of total dose. In Figure 6.20, the model

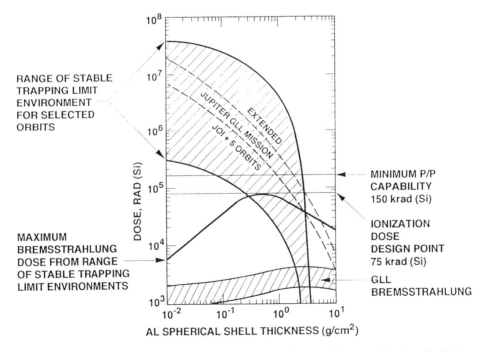

Figure 6.20. Comparison between the One-Year Dose from the Saturated Nuclear Radiation Environment for a Range of Earth Orbits and the Total Dose Expected for the *Galileo* Jupiter Mission after Five Orbits and after a Two-Year Extended Mission.

predictions for the total electron dose expected for *Galileo* as a function of different levels of aluminum shielding and two time intervals, closest encounter (with Jupiter) plus five orbits and after the two-year extended orbit (closest encounter plus 13 orbits), are graphed. These predictions are also compared with the range of dosages expected for a spacecraft in a circular orbit at various altitudes above the Earth in the saturated nuclear environment after one year. The typical shielding level on *Galileo* and a hardened DoD spacecraft is at least 3 g/cm^2. The hardness levels of the *Galileo* electronic components is 150 krads (Si), commensurate with this level of shielding. The use and impact of the model in determining the dose levels for the radiation hardness design of *Galileo* are obvious. It is also very interesting that these results are comparable with the one-year nuclear saturated environment for the Earth at 3 g/cm^2. The levels of dosage are remarkably similar and imply that radiation environments at least as severe as the saturated nuclear environment exist in space.

Another application of the Jupiter radiation model was in studying anomalies on *Voyager*. During the *Voyager I* flyby of Jupiter, 42 anomalies were observed. Leung et al. (1986) postulated that these anomalies were caused by arc discharges during the flyby. Although surface charging was ruled out as a possible cause, subsequent estimates of the total electron fluence indicated that there was a sufficient radiation environment of high-energy electrons present to cause internal electron charging. That is, it has been proposed that the $E > 100$-keV electron environment deposited sufficient charge in an external cable on *Voyager* to generate arc discharges that could cause power-on reset (POR) anomalies. To test this assumption, the total fluence of electrons at $E > 1$ MeV and $E > 10$ MeV and protons at $15 < E < 26$ MeV were computed as a function of time. The resulting normalized curves are plotted in Figure 6.21 relative to the cumulative sum of *Voyager* POR anomalies (cumulative is used because the charge buildup is a cumulative process). As this figure implies, internal electron-caused charging is a possible source of the *Voyager* anomalies in a temporal sense because the energetic electron fluence roughly follows the pattern of POR events. The major evidence for buried charge as the cause, however, came from an estimate of the charge deposited in each arc that would be necessary to cause the observed POR upsets. This indicated that the total $E > 100$-keV electron fluence (i.e., the minimum energy necessary to penetrate several mils of shielding and deposit charge) was very close to the actual level required to account for all 42 arcs. This is believed to be one of the earliest published examples of internal arc discharge occurring in space and demonstrates another application of radiation modeling. Subsequent ground testing of the *Galileo* vehicle supported the buried-charge hypothesis and revealed several potential sources of such arcing that led to its redesign.

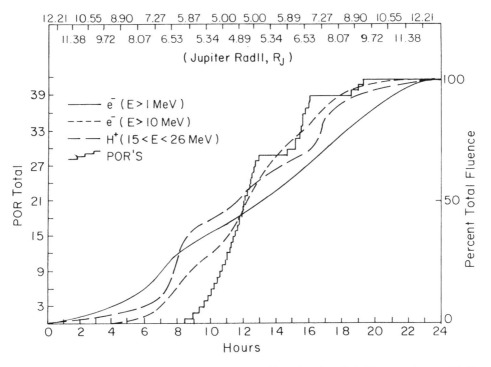

Figure 6.21. Correlation of *Voyager* POR Anomalies with the High-Energy ($E > 1$ MeV and $E > 10$ MeV) Electron Environment. Also Shown for Comparison is the High-Energy Proton Environment (15 MeV $< E < 26$ MeV). [From Leung et al. (1986).]

6.5.3 Heinrich Flux Estimates

The Heinrich flux curve is the starting point for most SEU calculations. Here, two practical examples of the effects of source, location, and shielding are presented. These calculations are based primarily on the results of Adams (1982). Consider the effects of magnetospheric shielding. In Figure 6.22, the Heinrich integral flux diagram for a spacecraft with 0.064 cm of aluminum shielding is graphed as a function of orbital inclination at an altitude of 400 km. A 90 percent worst-case flare and an AP trapped-proton environment are assumed. As noted by Adams (1982), below an LET of 0.5 MeV-cm²/mg, the population is dominated by the trapped protons. As the LET increases and the inclination increases, more and more flare particles can penetrate to the spacecraft. The steps correspond to the large drops in abundance beyond iron, nickel, bismuth, and uranium in the Adams-flare-model heavy ions. Figure 6.23, for the same environment, illustrates the variations produced by increasing the altitude for a 60° inclination orbit and for 0.064 cm of aluminum shielding. What variations exist are primarily below 0.5 MeV-cm²/mg and are due almost entirely to the trapped-proton environment. These variations go through a maximum between 4,000 and 6,000 km.

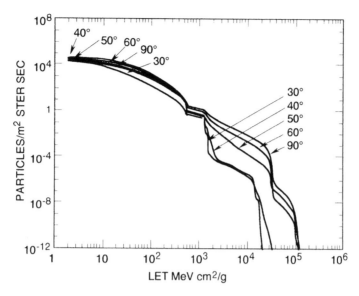

Figure 6.22. Integral LET Spectra Inside a Spacecraft (With 0.064-cm Aluminum Walls) in a 400-km Circular Orbit. The 90 Percent Worst Case Environment is Assumed in the Interplanetary Medium and the AP Trapped-Proton Environment at the Earth. The LET Spectra are for the Various Orbital Inclinations Indicated. From Adams (1982). Copyright © 1982 AIAA. Reprinted with Permission.

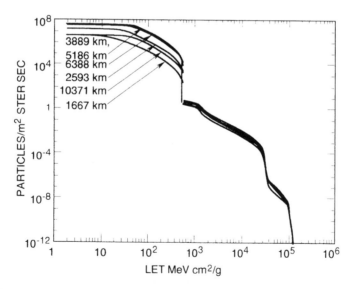

Figure 6.23. Integral LET Spectra Inside a Spacecraft with 0.064-cm Aluminum Walls that is in a Circular Orbit at a 60° Inclination. As in the Previous Figure, the 90 Percent Worst Case Environment is Assumed in the Interplanetary Medium and the AP Trapped-Proton Environment at the Earth. The LET Spectra are for the Various Altitudes as Shown. From Adams (1982). Copyright © 1982 AIAA. Reprinted with Permission.

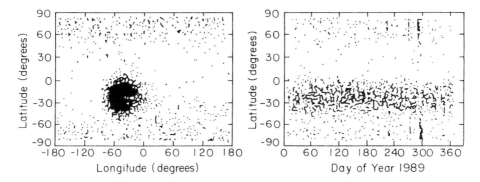

Figure 6.24. SEU Locations on the TMS4416 RAMs for UOSAT2. Used by Permission of DeWitt et al. (1993). Reprinted by Permission of Kluwer Academic Publishers.

First, the computations predict that, even for substantial levels of shielding, the solar-flare environment normally will dominate GCR rates in SEU calculations even at high-LET levels in interplanetary space. Indeed, these figures provide upper limits on the SEU environment expected at 1 AU (for no magnetospheric shielding). Second, given the increasing importance of solar-flares to the overall SEU environment, statistical techniques such as represented by the JPL solar-flare models need to be utilized for predicting SEU effects. Third, in Earth orbit, the picture is very complex because of geomagnetic shielding, but, at low LETs (i.e., LET <5–10 MeV-cm^2/mg), the trapped protons will normally dominate the environment. This is illustrated in Figure 6.24, which shows SEU locations plotted as a function of latitude and longitude. These were taken from the UOSAT2 satellite which was at a 700-km altitude and in a 98°-inclination circular orbit (Bourrieau, 1993). The map of SEU locations clearly corresponds to the South Atlantic anomaly (see Section 3.3.1) where the trapped-proton flux is the highest.

6.5.4 CRRES Results

CRRES was launched into an 18° inclination geosynchronous transfer orbit on July 25, 1990 (Mullen and Gussenhoven, 1993). The satellite functioned until October 12, 1991. One of the purposes of the satellite was to measure the dynamics of the near-Earth radiation environment and to measure the effects of the space environment on electronics, solar cells, and materials.

CRRES contained a space radiation dosimeter that measured the radiation dose by means of energy deposited from both electrons and protons behind four aluminum hemispheres of different thicknesses. The thinnest aluminum shielding was referred to as Dome 1, the next as Dome 2, and so forth. The characteristics of the detector and threshold were such that energy deposition between 50 keV and 1 MeV gave what was termed low linear energy transfer (LOLET) doses. Deposition between

Table 6.5. *Satellite effects of March 1991 storm*

MARECS-A	Satellite failure
GOES-7	Permanent power degradation (reduction in life of 2 years)
TDRSS	SEUs
INTELSAT	SEUs
CRRES	SEUs
NOAA-11	High spacecraft potential, degradation of sensors, cable discharges, loss of automatic attitude control
NORAD	Satellite drag, loss of 200 satellites

1 and 10 MeV gave high linear energy transfer (HILET) dose. Deposition above 40 MeV was termed very high linear energy transfer (VHLET) dose. The LOLET dose comes primarily from electrons, protons, and bremsstrahlung. The HILET dose is primarily from protons below 100–200 MeV. The VHLET dose comes from high-energy proton interactions inside or near the device (called nuclear star events) or from heavier cosmic rays. For Dome 1, the aluminum shielding was 0.57 g/cm^2, for Dome 2 it was 1.59 g/cm^2, for Dome 3 it was 3.14 g/cm^2, and for Dome 4 the shielding was 6.08 g/cm^2. *CRRES* also contained a microelectronics package (MEP) specifically designed to measure SEUs in electronic components.

CRRES was active during two large solar events that occurred between March 22, 1991 and May 1, 1991. One of the events was a solar flare larger than 99 percent of all measured flares. It began on March 22, 1991, and had a number of profound satellite effects that are documented in Table 6.5 (Sagalyn and Bowhill, 1993).

CRRES demonstrated that the influx of solar protons formed a second inner proton belt that remained stable up until the satellite ceased to function. This can be seen very clearly by a comparison of Figures 6.25 and 6.26. In Figure 6.25 the SEU rate is shown relative to the L shell for 35 proton-sensitive devices in the MEP. The data are for the period before the large solar events. The large peak at L = 1.5 coincides with the heart of the inner radiation belt. In Figure 6.26, the SEU rate is plotted for one month after the solar proton event of March 1991. The double-belt proton structure is evident in the double-peaked SEU frequency. For both plots, the near-constant SEU rate from GEO to the outer edges of the proton belts indicates that there is no sharp cutoff of the cosmic ray population that is causing the events. To distinguish between direct proton initiation of SEU as opposed to cosmic rays and nuclear events, the VHLET is plotted in Figure 6.27 relative to the L shell as well as to the SEU rate. The agreement is very good, indicating that in the inner

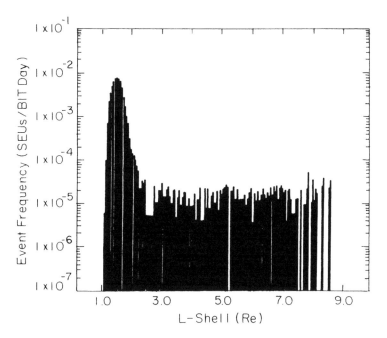

Figure 6.25. SEU Rate from CRRES Relative to L Shell for the Period July 25, 1990, to March 22, 1991. Used by Permission of DeWitt et al. (1993). Reprinted by Permission of Kluwer Academic Publishers.

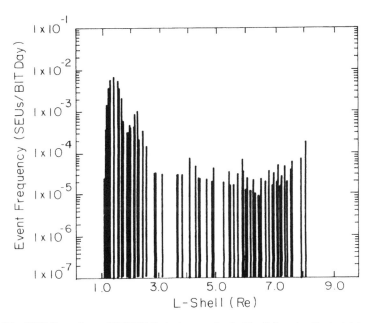

Figure 6.26. SEU Rate from CRRES Relative to the L Shell for the Period March 22, to April 22, 1991. Used with Permission of DeWitt et al. (1993). Reprinted by Permission of Kluwer Academic Publishers.

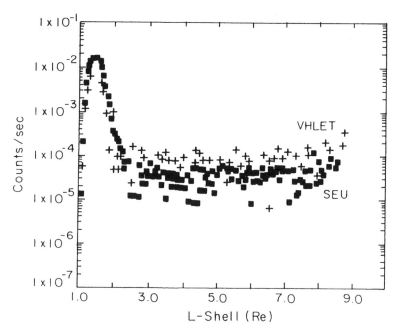

Figure 6.27. SEU Rate and VHLET Count Rate from CRRES Relative to the L Shell. Used with Permission of DeWitt et al. (1993). Reprinted with Permission of Kluwer Academic Publishers.

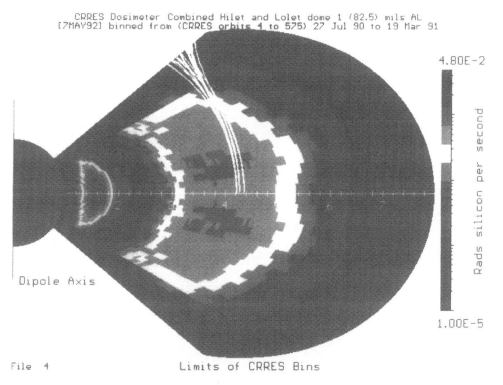

Figure 6.28. Average CRRES Dosimeter Combined HILET and LOLET Data Behind Dome 1. The GPS Orbit is Shown for Reference in the Upper Quadrant.

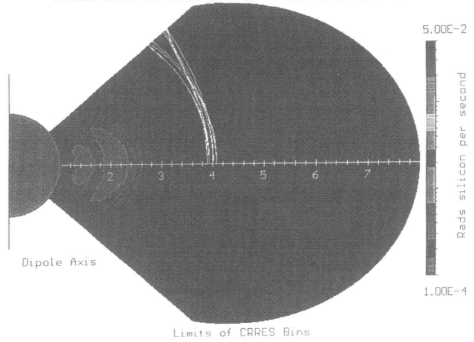

CRRES Dosimeter Combined Hilet and Lolet dome 4 (886.5) mils AL
[7MAY92] binned from (CRRES orbits 606 to 1067) 31 Mar 91 to 12 Oct 91

5.00E-2

Rads silicon per second

1.00E-4

Dipole Axis

Limits of CRRES Bins

Figure 6.29. High-Activity CRRES Dosimeter Combined HILET and LOLET Data Behind Dome 4. The GPS Orbit is Shown for Reference in the Upper Quadrant.

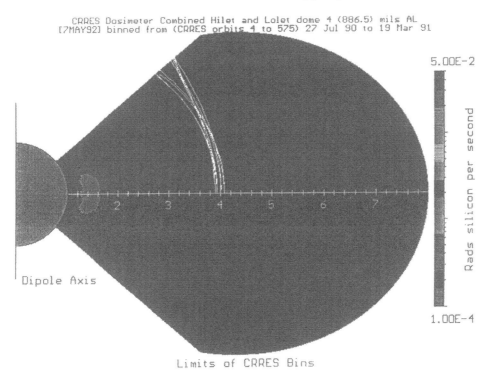

CRRES Dosimeter Combined Hilet and Lolet dome 4 (886.5) mils AL
[7MAY92] binned from (CRRES orbits 4 to 575) 27 Jul 90 to 19 Mar 91

5.00E-2

Rads silicon per second

1.00E-4

Dipole Axis

Limits of CRRES Bins

Figure 6.30. Low-Activity CRRES Dosimeter Combined HILET and LOLET Data Behind Dome 4. The GPS Orbit is Shown for Reference in the Upper Quadrant.

proton belt the SEUs are caused by nuclear star events rather than direct proton deposition in the chips.

Average combined HILET and LOLET doses are shown over the *CRRES* range behind Dome 1 in Figure 6.28. For reference, the GPS orbit is superimposed on the upper quadrant. This GPS orbit is at 55° inclination and at approximately 18,520 km. The orbit went through the outer radiation belt and accumulated an average radiation dose of 170 krads/year behind Dome 1 over the time that *CRRES* operated. In Figure 6.29, the dose behind Dome 4 is plotted over the stormy period of March 31, 1991, to October 12, 1991. Behind Dome 4, the shielding is so high that only 245 rads/year is accumulated. In Figure 6.29, the structure of the double inner belt is clearly seen. For comparison, in Figure 6.30, the data are shown for the quiet period before the flare of March 21, 1991. No double inner belt is seen and the average radiation dose behind Dome 4 for the GPS orbit is only 77 rads/year for this quiet period.

7

Particulate Interactions

7.1 Particle Impacts on Spacecraft

In this chapter, the effects of space particulates – hypervelocity impacts and scattering – are considered. Hypervelocity impacts, the primary effect of meteoroids and space debris, can be roughly divided into effects on single surfaces (namely, cratering or penetration of single surfaces), spall formation, and double-wall (or Whipple) shield penetration. A major consideration for each of these is the target. For example, if the target is an optical surface, then the damage induced on single surfaces is important. If the target is a tank, then penetration and/or failure of the tank is important and the characteristics of its contents become crucial. For electronic components inside a box, the size and distribution of the spall or spall/impactor products coming off the wall of the box are important. Examples of these factors are discussed below with emphasis on the practical considerations that must be taken into account in arriving at an effective and economical (in mass) protection system. A particularly important issue for meteoroid shielding design has arisen in recent years because of the need to have interplanetary spacecraft carrying nuclear RTGs use the Earth for gravitational assists. To provide adequate safety margins for these missions, which often require hypervelocity flybys of the Earth at distances of 300 km, mission planners must target the vehicle so that it will be extremely unlikely for a meteoroid impact on the spacecraft to lead to an Earth-impact trajectory. This typically translates into a careful estimate of the possibility of a propulsion-system rupture that could impart an unwanted change to the spacecraft flight path. To illustrate the methods involved in designing an optimal shielding system and in estimating the likelihood of failure, several practical examples from the *Galileo* mission are presented.

It is important to note that the particulate impact threat can be divided into three size ranges. Small-size particles with diameters below about 0.1 mm ($\sim 10^{-3}$ g) typically will not be of concern as a penetration threat but could lead to long-term erosion of surfaces through cratering. (*Note*: EVA activities might be threatened, however.) Above about 2 to 4 cm in diameter, it is impractical to provide shielding protection, and the only solution is avoidance. Between about 0.1 mm and 2 cm ($\sim 10^{-3}$ to 2 g), shielding can be provided for specific systems, although the mass penalty may be prohibitive. Thus, it is the effects of particles in this size range that need to be considered carefully for design purposes. Accurate knowledge of these effects will allow appropriate trade-offs between maximization of the protection and minimization of the shielding mass for spacecraft systems.

7.1.1 Hypervelocity Impacts and Shielding Theory

Aside from in-situ space data where the characteristics of the projectiles are poorly known, hypervelocity collision data are limited to the particle sizes and velocities that can be simulated in the laboratory. These in turn are generally limited to masses of <1 g and <7 km/s for large particles and to 1–70 km/s for $\sim$1-mg particles (usually glass or aluminum; e.g., Goller and Grun (1989) where velocities reached 40 km/s at 1.13×10^{-6} g). Impact data are invariably in terms of the effects on one or two surfaces (i.e., bumper shield and primary shield). Typical single-surface (infinite-plane) experimental values of the p/d (penetration depth/particle diameter) for normal incidence are presented in Figure 7.1 from Cour-Palais (1987) for aluminum particles from 50-μm to $\sim$1-cm diameter impacting aluminum. This type of information is used to establish a relationship between crater size/particle size and velocity for single surfaces.

When a hypervelocity particle strikes a shield, a shock propagates in both media. These two shocks are the controlling factors in the interaction. If the shield is thin

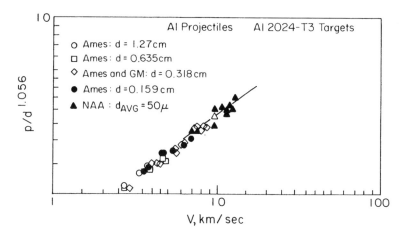

Figure 7.1. Penetration Depth/Particle Diameter, $p/d^{1.056}$, Relative to Impact Velocity for a Range of Particle Diameters. (Cour-Palais, 1987).

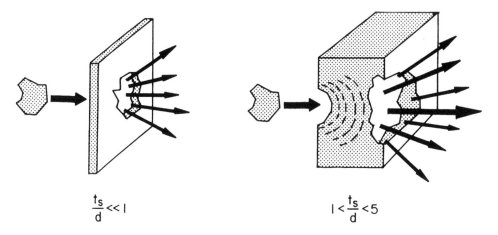

$$\frac{t_s}{d} \ll 1 \qquad\qquad 1 < \frac{t_s}{d} < 5$$

Figure 7.2. Schematic of Effects of Hypervelocity Impact for $t_s/d \ll 1$ and $t_s/d \gg 1$.

relative to the diameter of the particle (assumed spherical) (i.e., $t_s/d \ll 1$ where t_s is the shield thickness and d is the particle diameter), then the shock that propagates into the shield can reflect from the rear of the shield and overtake the compression wave in the impacting particle. This disrupts the compression wave so that the particle does not have time to get very hot. Consequently, the shield breaks into solid fragments or the particle cuts out its shape ("cookie cutter") and the impacting particle itself is not disrupted. As the ratio t_s/d increases, it reaches a value where the reflected shock just overtakes the compression wave in the particle as it reaches the end of the particle. By this time, depending on the peak shock pressure generated at the instant of impact, the particle will be heated to its melting or vaporization point. The interaction of the two waves gives a radial velocity to the melted particle, and both particle and shield are destroyed. As the ratio t_s/d continues to increase ($t_s/d \gg 1$), the reflected shock does not have time to overtake the compression wave in the particle. The impacting particle thus melts or vaporizes and moves cohesively into the shield. Material from the shield may be thrown from the back of the shield. This is known as the spall of the shield. If the shield has a low melting point, then the spalled material is likely to be in a melted state. The two limits are illustrated in Figure 7.2.

Another way to look at this range of interactions at an elementary physical level is as follows: for impacts where the projectile's mass per unit area is much larger than the shield's and its impact energy is much greater than that needed to penetrate the target (and the energy transfer from the impact is not high enough to damage the projectile), the impactor can act like a cookie cutter. It punches out a crater roughly proportional to its cross section in passing through the shield. In this case, the effects of the shield on the impactor are primarily a momentum reduction proportional to the mass of the shield material and the velocity of the projectile. At the other extreme, for cases where the projectile does not penetrate the shield and the velocity is so high that it can be destroyed, the crater created should be proportional to the energy of the impact. Although the actual crater size and shape are dependent on many different physical parameters, the crude proportionality of the mass removed from the crater to either the momentum or the energy of the projectile is a starting assumption for the shielding models to be presented. In particular, the data in Figure 7.1 are fitted by relationships that are proportional roughly to the energy of the impacting particle (Cour-Palais, 1987; or Swift, Bamford, and Chen, 1982). First, the theoretical formulation of Swift et al. (1982) is considered. Assuming that the craters are hemispherical and that all of the kinetic energy goes into melting the impacting particle, it gives,

$$\frac{4}{3}\pi\left(\frac{d_m}{2}\right)^3\frac{\rho_m V_m^2}{2} = \frac{1}{2}\frac{4}{3}\pi\left(\frac{d_c}{2}\right)^3 H_f, \tag{7.1}$$

where d_c is the crater diameter, d_m is the projectile diameter, ρ_m is the density of the projectile, V_m is the velocity of impact, and H_f is the heat of fusion. Because a given particle may also vaporize or not melt smoothly, a general relation between d_c and d_m can be written as

$$\frac{d_c}{d_m} = \left(\frac{\rho_m V_m^2/2}{E_0}\right)^{1/3} \tag{7.2}$$

for some constant E_0. For the ideal case in Eq. (7.1), $E_0 = H_f/2$. For a low-density particle striking aluminum, $E_0 = 10^9$ J/m^3. Another formula in widespread use is that of Cour-Palais. It is based on empirical fits to data and gives (in the same format)

$$d_c = C d_m^{1.056} \rho_m^{0.519} V_m^{2/3}, \tag{7.3}$$

where d_c is the diameter of crater in centimeters, d_m is the diameter of projectile in centimeters, ρ_m is the density of projectile in grams per cubic centimeter, V_m is the velocity of impact in kilometers per second, and C is a constant that is characteristic of the target material and its temperature. The Swift and Cour-Palais formulas are only a small subset of the many different cratering formulas that have been used to fit hypervelocity data.

Note that the distribution of particles at small sizes is often determined by measuring the crater size and distribution on samples of surfaces from returned spacecraft. The crater size is then related to the mass or particle diameter (if given the density) by equations such as those above. Although it is important to have the right relationship for the mass/crater size and to work from this relationship back to the particle distribution function, it is sometimes practical and more accurate to use

Table 7.1. *Number of holes and craters, by size, for exterior surfaces on* Solar Max

		Impact feature diameter (μm)							
	Total	≥ 40	≥ 60	≥ 80	≥ 100	≥ 120	≥ 160	≥ 200	≥ 300
Holes	492	487	447	406	372	300	217	164	73
Craters	1,416	1,147	480	282	180	103	42	19	4
Both	1,908	1,634	927	688	552	103	259	183	77

the original crater distributions. Here, as an example, the "raw" crater distributions as actually observed on the *Solar Max* mission are presented in Table 7.1 (Zook, 1987). These size distributions for an aluminum surface provide estimates of the actual number and distribution of pits to be expected in this orbit – likely a much more accurate estimate of small-particle surface effects than any modeled values. In particular, they include both the interplanetary and debris particles most likely to be encountered. Such tables (particularly for the recently published *LDEF* data), with appropriate geometric shielding calculations, are used to determine the expected surface distribution of hypervelocity craters from which long-term pitting and erosion effects can be estimated in LEO. Unfortunately, similar data on surface erosion do not exist for interplanetary space (except for lunar samples and a *Surveyor* scoop).

At larger sizes and for the shielding of internal components, estimates are required of the effects of the spall/secondary projectile products that will be produced by the impacting particle. It is these byproducts of the collision that will more often cause the damage than the particle itself. Two cases need to be considered. First, the single-surface-impact spall and, second, the spall from a double or "Whipple" meteoroid shield – the currently preferred shielding technique for many spacecraft. As discussed, different formulas relating spall/crater size to the impacting particle are derived, depending on whether the interaction is assumed to be a momentum or energy interchange. In addition, the conditions behind the shield surface are very important in determining the ultimate effects of the impact – that is, is the primary shield surface backed by a pressurized gas or liquid or is it a vacuum? As a first example, only the simple case of a single-surface unpressurized vessel is considered. In that case, the work of Cour-Palais (1987) and Swift et al. (1982) are the principal theoretical models. Both Cour-Palais and Swift et al. have calculated the critical shield thickness for which penetration occurs, based on the previous formulas for crater depth. The model of Swift et al. assumes that critical penetration occurs when the shield thickness is equal to the depth of the crater associated with the spall plus the depth of the front surface crater (i.e., a connecting neck is just formed between the two craters). The work of Cour-Palais assumes the creation of a "minimum" spall (i.e., where the projectile is stopped in the single-surface shield just as it creates a detached spall) and provides an estimate of the maximum level of protection possible for a single-surface shield. These two formulations are representative of the range of models to be found in the literature. The respective formulae are

Swift/Bamford/Chen:

$$t_c = \frac{3}{2}\left(\frac{d_c}{2}\right) = \left(\frac{3d_m}{4}\right)\left[\rho_m V_m^2/(2E_0)\right]^{1/3}, \tag{7.4}$$

Cour-Palais:

$$t_c = C_o \rho_m^{1/6} m^{0.352} V_m^{0.857}, \tag{7.5}$$

where t_c is the critical thickness of shield for which penetration will occur for particles equal to or greater than the critical size (in centimeters for the formula of Cour-Palais), C_o is a constant for the shield material $= 0.351$ (for Al), and m is the impacting mass in grams.

The results, namely the critical penetration mass for a given impact velocity, are presented in Figure 7.3 for an approximately 0.1-cm-thick shield of aluminum (equivalent to the single-surface protection levels on the *Galileo* spacecraft). Two cases are plotted: one for a density of 0.5 g/cm^3 (equivalent to cometary meteoroids) and one for 3.5 g/cm^3 (equivalent to the asteroidal meteoroids and space debris). The spall size would be roughly that of the crater size.

Although single-surface meteoroid shields are adequate for many purposes, mass considerations usually prevent sufficient protection thicknesses. It has been found empirically that for a low-density particle impacting aluminum, the critical thickness for no penetration and no spall is at least five times the particle diameter. Hence, for even small impacting particles, a single-surface shield may be too massive. This suggests the idea of a bumper shield. The concept of using a bumper shield or surface in front of the main surface was first suggested by Whipple (1947). The basic idea of a bumper shield is that the bumper will disrupt the incoming particle and the spacing between the bumper and the second surface will allow the disrupted cloud of particles, liquid, and gas to expand, dissipating the energy associated with the incoming particle over a larger area. In principle, the two shields, if properly sized, may have a significantly smaller combined mass than an equivalent single-surface shield. Numerous hypervelocity tests have indeed revealed the value of this concept. The multisurface shield, however, must be designed for a specific mass/velocity range of particles – outside this range (i.e., at too low or too high a thickness for the front surface for a given velocity), the shield will not function properly and may even enhance the danger. The physics behind the second surface shield and the problems associated with it are discussed below.

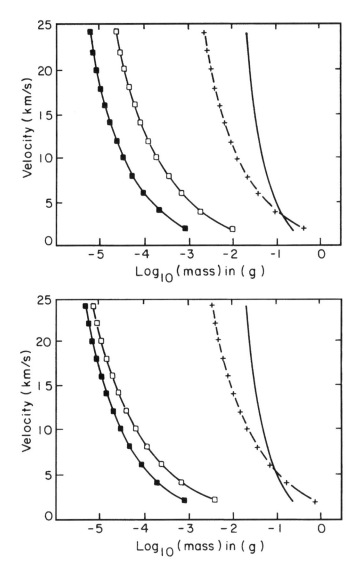

Figure 7.3. Plots of Critical Penetration Mass as a Function of Impact Velocity for Cometary Meteroids (Top Plot) and for Asteroidal Meteorites and Space Debris (Bottom Plot). Black Squares are for the Swift Model, Open Squares are for the Cour-Palais Model, Crosses are for a Nonoptimum Second Surface Shield, and the Line is for an Optimum Second Surface Shield.

First, consider the bumper shield. When the shield is sufficiently thick or the particle is moving fast enough, the incident particle will be completely disrupted – typically melting or vaporizing. If, on the other hand, the shield is too thin or the velocity of the particle is too slow, the particle may breach the shield largely intact with the mass of the extra shield material added. Similarly, if the shield is not too

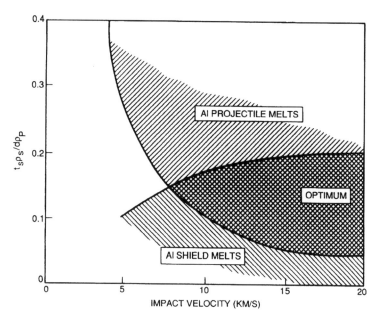

Figure 7.4. Optimum Range for Shield Design.

thick and the incident particle is moving fast enough, the shield material in the path of the particle will be liquified or vaporized. If, however, the shield is too thick or the velocity is too low, the shield material will either spall or come off in large pieces – in this way the bumper shield can actually enhance the danger from a meteoroid. An example of these effects is presented in Figure 7.4. The shaded region corresponds to the optimum region for a second surface shield. This optimum region is that range of mass, thickness, and velocity for which both the particle and the shield debris come off the back of the bumper as a liquid or gas. The objective of good shield design is to maximize this region for the particle mass range of most concern to the particular mission.

In addition to the bumper shield, the second surface also must be sized with care. Aside from structural considerations (i.e., the tank walls must be thick enough to contain the gas or liquid that they hold), the second surface must be able to withstand the incident pressure pulse or debris cloud from the bumper shield. Given the material characteristics of the second surface, two parameters determine its design – the thickness and the spacing. The spacing allows the cloud of gas or liquid to expand and dissipate (too far apart, however, and the structure becomes cumbersome or the liquid and gas begin to condense). As in the case of the single surface, the second surface must be thick enough to withstand spallation from either the liquid and the gas or from the debris. These considerations form the basis of the mathematical formulation of the second surface shield.

Cour-Palais details the steps actually required to design an optimum second-surface meteoroid shield. The first step, as outlined above, is to pick the thickness of the bumper shield so that, for the size and velocity range to be protected against, the impact conditions fall in the "optimum" range or close to it. That is, the ratio of the thickness of the bumper shield to the particle diameter should be between the curves for "projectile melts" and "shield melts." These are given approximately by

$$(3.16/V_m^{1.5})(\rho_m/\rho_l) < t_s/d < (0.24 - 0.7/V_m)(\rho_m/\rho_l), \qquad (7.6)$$

where ρ_l is the bumper shield density (g/cm^3) and t_s is the bumper shield thickness in centimeters. The factor (ρ_m/ρ_l) allows scaling of the results shown in Figure 7.4 to other materials.

The second step is, given the shield spacing, to determine the thickness of the second surface or, given the thickness of the second surface, to determine the shield spacing. The optimum solution is defined as the condition that the particle and shield debris are either completely molten or vaporize. These quantities have been found empirically for the optimum solution to be related by

$$t_2 = (C_l m V_m)/S^2, \qquad (7.7)$$

where t_2 is the thickness of the second surface shield (cm), C_l is a material constant $=$ 41.5 (for Al), m is the mass of impacting particle (g), and S is the shield spacing (cm).

Similarly, for the nonoptimum solution (defined according to Cour-Palais to be on both sides of the optimum range, i.e., either the impacting particle gets through as a solid particle or some of the bumper shield material comes off as solid fragments),

$$t_2 = \left(C_2 d_m V_m \rho_m^{1/3} r_l^{1/6}\right)/\left[(S/d_m)^{.4}(C_3^{.5})\right] \qquad (7.8)$$

where C_2 is a material constant $=$ 33.6 (for Al), C_3 is another material constant $=$ 42,000 (for Al), and r_l is the density of shield $=$ 2.8 g/cm^3 (for Al).

For both cases, Cour-Palais suggests that the formulas are only valid for $(S/d_m) <$ 100. These equations can be used to determine either the shield thicknesses or, by inverting the formulas, the critical penetrating mass for a given shield configuration and velocity. Such an example of a mass/velocity curve for characteristics similar to the *Galileo* RPM meteoroid shield are presented in Figure 7.3 for the optimum and the nonoptimum solutions for both low-density 0.5-g/cm^3 (cometary) and high-density 3.5-g/cm^3 (asteroidal or debris) projectiles. Note that the two curves cross (this crossover is outside the optimum range, so the optimum solution is not valid at that point).

As another example of this procedure, consider the case of the *Galileo* helium pressurant tanks. The titanium helium tanks are inside the *Galileo* thermal/meteoroid protection blanket. For the purposes of this calculation, it is assumed that for the blanket the incident meteoroids are in the optimum range. The thickness of the tank wall is taken to be 0.35 cm and the constant $C_1 = 20.8$ for titanium. For Eq. (7.7), this gives in terms of the particle mass, velocity, and shield spacing,

$$m = 0.0168 S^2 / V_m. \tag{7.9}$$

This relationship and the meteoroid flux models are combined in a subsequent section to estimate the flux capable of penetrating the *Galileo* helium tanks.

To summarize the Whipple shield example, the range of particles to be protected against is selected first. Then, given the appropriate material characteristics and possible shield configurations, the bumper shield and primary shield thicknesses and spacing distance are determined. This is an optimization process and requires various trade-offs. The characteristics of the second or primary shield may complicate the process in realistic cases. In particular, for the *Galileo* spacecraft, where protection of liquid filled tanks with thermal blankets was desired (i.e., the bumper shield was not a continuous piece of metal but made up of thin layers of different materials), it was difficult to determine the critical mass/velocity curve. The resulting mass/velocity curves, when combined with the models of the flux of debris and meteoroids, are then used to determine the likelihood of a spacecraft being damaged in the space environment by a debris or meteoroid impact.

To conclude this section, the following recommendations are made:

(1) Where possible, in-situ measurements of the pit size distributions (for the desired orbit), rather than model fluences and impact models, should be used to estimate the number and size distribution of small impact craters to be expected on typical spacecraft surfaces. Such data are now available for ∼500-km altitude orbits around the Earth.

(2) For the protection of internal components, estimates using the particle distributions and critical mass penetration models are appropriate for first-order estimates in most cases. These estimates, however, are actually valid for only a limited range of conditions, and adequate margins always should be included.

(3) The effects of Whipple shield are to spread out the debris over a large area, depending on the bumper shield spacing, while reducing the combined weight of the shielding below that of an equivalent single-surface shield. As a special consideration, if redundant electronic circuits are used as a means to ensure survivability, then they should be spaced far enough apart to ensure that this larger debris cloud will not destroy both sets at one time.

7.1.1.1 A Real Case: Multisurface Shield for Propellant Tanks

The actual shielding design for propellant tanks is much more complex than the simple models presented. Although many situations can be modeled in terms of these simple one- or two-surface shield configurations, propellant tanks are normally covered by a multilayered thermal blanket. Adding further complexity is the fact that the propellant tanks may contain pressurized liquids or gas. As a result, the penetration processes leading to tank failure due to meteoroid impact are somewhat different from those described. Even so, the basic concepts introduced previously of cratering/spalling and of a bumper shield breaking up the meteoroid are all still valid. Swift and Bamford in two studies have extended existing theories to include complex tank configurations. Here, only the gross features of their theory are summarized. Even so, the physical relationship between the three models should be evident.

Consider the *Galileo* propellant tanks. These tanks are covered by a multilayer blanket that serves both for thermal control and for meteoroid protection. The blanket is made of a number of different materials, some of which are in the form of netting, and the layers have small but variable spacing between them. The layers thus serve as numerous thin bumper shields – so thin, however, that the meteoroid simply "punches out" its shape as it passes through. Swift and Bamford have assumed that each of these multiple shield encounters is a small shock wave imposed on the impacting body. They postulate that the result of all these impacts is, in the case of cometary meteoroids, the crushup and vaporization of the material. Because of the much higher density structure of the asteroidal meteoroids, they will most likely crumble and not be completely vaporized. For the velocity and mass range of interest then, the result of the thermal/meteor protection blanket is to fundamentally alter the state of the impacting particles – much like the bumper shield. Note, however, that if the impact is at very low velocities ($V_m \ll 10$ km/s), the asteroidal meteoroids may come through the shield largely intact – this threat is considerably less severe than high-velocity impacts, but must be handled separately.

The second key feature of the Swift/Bamford model of propellant tank failure calculation is the fact that the tanks are filled with a liquid. This problem required the development of a new theory of tank rupture. It was found that several different failure mechanisms can occur, depending on the characteristics of the debris cloud. If the impact fragments (for asteroidal meteoroids of large sizes or at very slow speeds) are large enough, pin-holing/spalling, as in the case of a single surface, can occur. If the debris cloud has sufficient energy, it can cause punchout. A sufficiently large punchout hole can produce a propagating crack that, when it reaches the tank ullage, can cause the tank to shatter. The conditions were computed under which a fluid-filled tank would fail due to the extended (in area) pressure pulse of the debris cloud. These results then determined the details of when no spall would occur, when an acceptable spall would occur (i.e., no failure), and when failure would occur. The limits on punctures were also estimated. The results of this analysis for a representative configuration (the *Galileo* mass per unit area ratio and a 10-cm or 15-cm spacing) are presented in Figure 7.5. As would be expected, there are

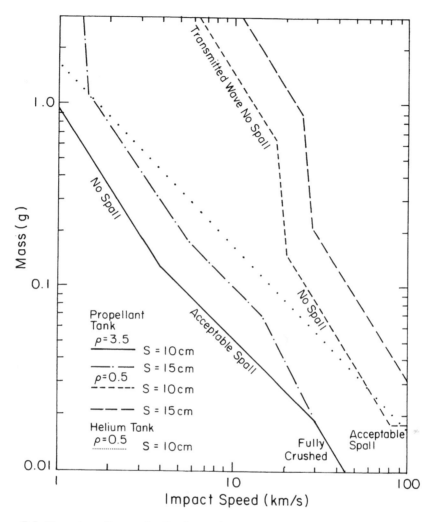

Figure 7.5. Boundary Curves for Regions of Tank Failure and Nonfailure. (Garrett and Petrasek, 1988).

significant differences between the cometary and asteroidal meteoroid penetration curves in this diagram.

7.1.1.2 Meteoroid Failure Probability Calculations

The preceding sections have concentrated on the development of models of the meteoroid integral flux above a given mass, of the relative impact velocity, and of the critical mass as a function of impact parameters (velocity, density, shield spacing, and so forth). These equations are combined in this section, along with an estimate of the appropriate sensitive area, to calculate the probability of an impact failure of a particular spacecraft system, the propellant tanks on the *Galileo* mission.

In statistical terms, the probability of X impacts on a spacecraft is given by:

$$P(X, t) = \frac{(f_p A_s t)^X \exp(-f_p A_s t)}{X!} \qquad (7.10)$$

The probability of one or more impacts occurring is, by experience, very low. Thus, to a high degree of accuracy, the probability of one or more impacts of a meteoroid on a tank is given by subtracting the probability of no hit $[P(0) \simeq 1 - f_p(t)A_s t + \cdots]$ from 1:

$$P(X > 0, t) = 1 - P(0) \simeq f_p(t)A_s t \qquad (7.11)$$

or

$$P(X > 0, T) = \int_0^T f_p A_s dt, \qquad (7.12)$$

where t is a small time interval, T is time (mission duration), f_p is the penetrating meteoroid flux as a function of time, and A_s is the equivalent sensitive area (for *Galileo*, 8.6 m^2).

The procedure for estimating the probability of meteoroid impact is as follows:

(1) Set up the function(s) for determining the critical penetration mass. Typically this is done by developing a relationship between the mass, impact velocity, and/or other shielding parameters (such as the spacing).
(2) Determine the spacecraft orbital position, velocity, and direction.
(3) From the orbital characteristics, estimate the average relative velocity and, for the cometary model, the δ factor.
(4) From the average relative velocity, assumed to be the impact velocity (*Note*: It may be necessary to correct the impact velocity by the correction factor δ. This will be discussed in Section 7.1.1.5), and any other required parameters, determine the critical penetration mass (defined as the smallest mass capable of causing failure).
(5) Calculate the integral meteoroid number density at the critical mass.
(6) Compute the product of the number density and the relative velocity (corrected by δ if necessary). One-fourth of this is the flux, f_p.
(7) Sum the product of the flux and time interval to obtain the fluence as a function of mission time.
(8) Calculate the product of the sensitive area and the penetrating fluence to obtain the probability of impact.

This process, with minimal modification, is used for most meteoroid failure/ shielding calculations. For illustration, examples for the propellant and helium pressurant tanks on *Galileo* for the three meteoroid environments are presented.

7.1.1.3 Helium Tanks

As a simple example of the calculation of impact probabilities, consider the shielding configuration for the two *Galileo* helium tanks. The tanks are filled with helium gas and, to first order, their configuration approximates a Whipple shield. The critical penetrating mass is proportional to the inverse of the relative or impact velocity, a function of the orbit, and the square of the spacing [Eq. (7.9)] for both asteroidal and cometary meteoroids. The spacing, on the other hand, is a complex function of the shielding geometry. The shield is approximated by a half-cylinder of diameter 48 cm and length 96 cm; the helium tank is approximated by a 19-cm-radius sphere. The flux at a given point on the helium tanks varies with direction as the spacing varies. This variation is dependent on the model assumed for how the particles are affected by passage through the bumper shield. As a conservative estimate, it is assumed that any meteoroid hitting the shield, no matter from what direction, will produce a debris cloud that will expand uniformly from the point of impact (i.e., there is no net velocity along the path of the impacting particle following impact). Because of the geometric configuration (the helium tanks are spherical), the debris wave will first encounter (and presumably cause to fail) the tank along a radial vector from the center of the tank to the surface of the shield at the impact point. From Figure 7.6, the separation distance from a given point on the tank to the shield is then given by:

$$S = XX - r \qquad (7.13)$$

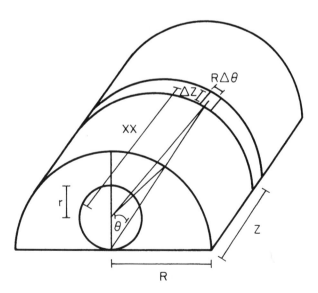

Figure 7.6. Schematic of *Galileo* Helium Tank. (Garrett and Petrasek, 1988).

where $XX = (X^2 + Z^2)^{1/2}$, $X = (R^2 + r^2 - 2rR\cos\theta)^{1/2}$, R = radius of the cylinder, r = radius of the helium tank, θ = angle between normal to the rear surface of cylinder at the intersection of the tank with the rear cylinder surface and the impact point, and Z = distance along the cylinder.

It was found empirically that the fluence to a helium tank protected by a uniformly spaced shield (spherically symmetric) of spacing "A" could be related to a similar shield of spacing "B" by the simple ratio:

$$F_p(A) = F_p(B)(B/A)^W, \tag{7.14}$$

where $W = 2Q$ and $Q = 0.84(1.2)$ for asteroids (comets).

Equation (7.14) follows because $F_p \propto m^{-Q}$ (from the meteoroid flux models) and $m \propto S^2$. This implies, for the cylindrical symmetry of the bumper shield, that the probability of impacts is given by (assuming $B = 10$ cm) the double summation:

$$P = F_p(10) \sum_j \sum_i (dZ R d\theta)(10/S_{ij})^W \tag{7.15}$$

where dZ = the increment size for Z (-48 cm $< Z < 48$ cm), j = steps in Z ($j = n$, $Z_n = n\,dZ$), $d\theta$ = the increment in angle θ ($-90° < \theta < 90°$), i = steps in θ ($i = n$, $\theta_n = n\,d\theta$), and S_{ij} = the distance from location (i, j) to the tank surface.

The quantity $(dZ R d\theta)$ is the small element of surface area on the cylinder where the impact takes place. Thus P is the sum of all of these small surface areas times the fluence scaled by their separation relative to 10 cm. This is a conservative estimate because the total fluence over 2π steradians at each small surface area is assumed to contribute. In reality, the fluence from a much smaller solid angle would contribute. The probability derived in this fashion gives the values listed in Table 7.2 for the failure of the two helium tanks for the *Galileo* mission. These values are believed to be about a factor of two to four too high, based on other estimation techniques. In particular, treating the problem in a manner similar to a radiation shielding problem gives values approximately one-fourth those of Table 7.2.

7.1.1.4 Propellant Tanks

Although the estimation procedure for the probability of a propellant tank being punctured by a meteoroid closely resembles that of the helium tanks, there is a major difference in the way that the critical penetration mass is calculated. As has been discussed, the mass/velocity relationship for multishield/liquid-filled tanks (Figure 7.5) yielded complex boundary curves for regions of tank failure/nonfailure. As an illustration, the near-vertical segment between velocities of 18 and 21 km/s in Figure 7.5 is the boundary where the size of the predicted tank spall would be greater than the thickness of the tank wall. Too large a spall means no spall, because

Table 7.2. *Integral meteoroid impact probabilities for launch through second Earth flyby: October 9, 1989 to December 8, 1992*

	Propellant tanks	Pressurant tanks
Number	4	2
Radius (m)	0.38	0.19
Effective shield thickness (g/cm^2)	0.0564	0.0564
Area (and spacing) (m^2)	0.86 @ 10 cm 7.74 @ 15 cm	(See Figure 7.6)
Median impact speed (approximate)		
Cometary (km/s)	20	20
Asteroid (km/s)	12.5	12.5
Median penetration mass (g)		
Cometary (10 cm)	0.3	0.09
(15 cm)	1.3	—
Asteroid (10 cm)	0.04	0.13
(15 cm)	0.08	—
Probability of failure		
Cometary	6.07×10^{-6}	6.12×10^{-6}
Asteroid	4.37×10^{-4}	2.46×10^{-6}

this occurs when the duration of the pressure pulse due to the impinging particle is longer than twice the time for the shock wave to cross the tank wall. Thus failure occurs for those velocities for which the spall is thinner than the wall (toward higher velocities) and survival for all velocities that are lower (to the left) than the critical value. The curves at even higher velocities correspond to boundaries separating where the spall is too small to affect failure or of sufficient size to just cause failure.

The behavior of the propellant tank shield system depends critically on the density of the impacting particles, the thickness (actually area-to-mass ratio) of the shield, and the spacing. Figure 7.7 presents the results for the failure of the propellant tanks due to impact with cometary meteoroids. Figure 7.8 presents the results for the failure of the tanks due to impact with asteroidal meteorites. A composite shield for *Galileo* is assumed to have 10 percent of the area at 10-cm spacing and 90 percent at 15 cm. Increasing the spacing increases the critical penetrating mass at a given velocity and therefore decreases the fluence and probability of impact. Further, as would be anticipated, the probability of failure due to an asteroid hit rises in the asteroid belts and approaches zero near the Earth and Venus (because it is the integral of the probability that is plotted, this means that the probability rises in the asteroid belts and is constant near the Earth). In contrast, the probability for cometary impact rises fairly uniformly during the first three years (the $R^{-1/2}$ velocity dependence is evident, however).

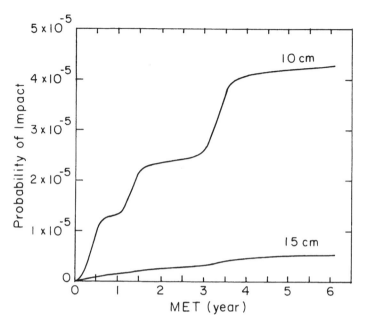

Figure 7.7. Probability of Failure for Propellant Tanks Under Impact from Cometary Meteoroids. (Garrett and Petrasek, 1988).

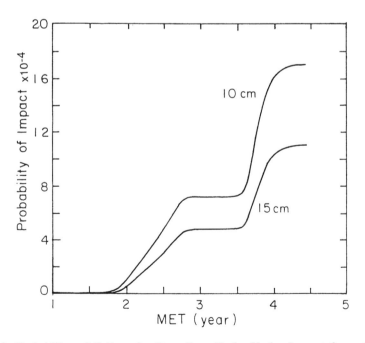

Figure 7.8. Probability of Failure for Propellant Tanks Under Impact from Asteroidal Meteoroids. (Garrett and Petrasek, 1988).

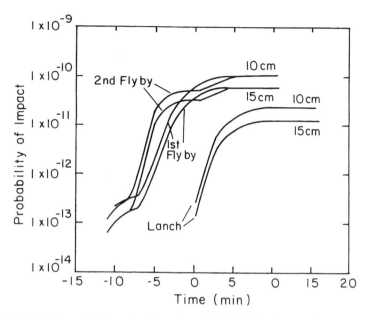

Figure 7.9. Probability of Failure for Propellant Tanks Under Impact from Orbital Debris. (Garrett and Petrasek, 1988).

The final population threatening the propellant tanks is the Earth debris environment. This debris environment represents a population in between that of the asteroidal and cometary environments but with many features of the asteroidal component. Like the asteroids, the debris particles are concentrated in a specific region – near the Earth. The debris density of 2.8 g/cm^3 is closer to that of the asteroidal than to the cometary particles. Therefore, as a conservative estimate it is assumed that the debris particles have the same mass/velocity relationship as the asteroids. This gives the impact probability curves in Figure 7.9. The debris can be an important threat but, as should be apparent from these curves, the short temporal duration of that threat limits its significance for Earth-gravity-assist flyby trajectories.

Throughout these calculations, the average relative impact velocity for the cometary component has been calculated as a fixed velocity of 20 km/s. To the accuracy of the calculations carried out, this is a reasonable assumption. In reality, as outlined at the outset, the velocity of impact actually follows a Gaussian-like distribution. The correction factor called δ is intended to account for this behavior. Section 7.1.1.5 completes the analysis of the *Galileo* tanks based on the NASA meteoroid models by defining this function and estimating its effects.

7.1.1.5 The δ Correction Factor

In the calculations of the probability of meteoroid penetration, the relative impact velocity (more properly, speed) should be weighted over the actual velocity distribution in terms of speed and angle relative to the spacecraft frame of reference. A common method adopted in the NASA monographs is to increase the average of the relative velocity, $\overline{V}_r$, by a δ factor, which is a function of angle and spacecraft velocity relative to a circular orbit at that point in space. That is,

$$\overline{V_r^n} = \delta^{n(n-1)/2}\overline{V}_r^n. \tag{7.16}$$

In terms of the velocity distribution, $\overline{V_r^n}$ can be expressed as

$$\overline{V_r^n} = \int W(V_r)V_r^n dV_r \Big/ \int W(V_r)dV_r, \tag{7.17}$$

where $W(V_r)$ is the phase-space velocity distribution as a function of speed V_r. The phase-space distribution can be approximately fitted by

$$W = K\exp\{-[(V_r - \overline{V}_r)/s]^2\}, \tag{7.18}$$

where s and K are constants, and s has the significance of the standard deviation of the relative velocity. Note that this distribution function is similar to the Maxwellian speed distribution defined in Eq. (2.25) and is basically identical to $N(V)$ discussed in Section 3.5.2.1. This means that the integrations worked out in Chapter 2 can be carried over to the particulate distribution function.

The distribution function in Eq. (7.18) can be integrated to give

$$\begin{aligned} \overline{V_r^1} &= \overline{V}_r \\ \overline{V_r^2} &= \overline{V}_r^2 + s^2 \\ \overline{V_r^3} &= \overline{V}_r^3 + 3s^2\overline{V}_r. \end{aligned} \tag{7.19}$$

With the definition, $\delta = 1 + (s/\overline{V}_r)^2$, and the assumption $s/\overline{V}_r \ll 1$ then for $\delta^{n(n-1)/2}$

$$\begin{aligned} n &= 1 & \delta^0 &= 1 \\ n &= 2 & \delta^1 &= 1 + (s/\overline{V}_r)^2 \\ n &= 3 & \delta^3 &\approx 1 + 3(s/\overline{V}_r)^2 \\ n &= 4 & \delta^6 &\approx 1 + 6(s/\overline{V}_r)^2. \end{aligned} \tag{7.20}$$

Therefore,

$$\overline{V_r^n} \approx \overline{V}_r^n \delta^{n(n-1)/2}, \tag{7.21}$$

and δ is now definable in terms of $\overline{V}_r$ and s. A numerical fit to the NASA cometary model yields the following approximation (all units are km/s) to s:

$$\begin{aligned} s &= 10.9\sigma + 0.87 & \sigma &< 1.159 \\ s &= 13.5 & \sigma &> 1.159, \tag{7.22} \end{aligned}$$

where σ is the ratio of the cometary heliocentric velocity to the velocity on a circular orbit around the Sun at the same distance.

Given δ, any estimate of a function in the NASA model is corrected by δ to the appropriate power. Consider a simple shielding case where $m \propto V_r^n$ (such as for the helium tanks), the calculation of the spatial density δ correction factor is as follows:

$$m \propto V_r^{-2}$$

and

$$\rho \propto m^{-1.2}.$$

Therefore,

$$\overline{\rho} \propto \overline{V_r^{2.4}} \sim \delta^{1.68}\overline{V}_r^{2.4},$$

and the correction factor for ρ is $\delta^{1.68}$. The maximum value of δ is 1.42, so the maximum correction factor is 1.8. The correction factor for $(\overline{V^{-1}})^{-1}$ is $\delta^{-1} = 0.7$, so that the flux correction factor is 1.27 (i.e., $F = \frac{1}{4}\overline{\rho}(\overline{V_r^{-1}})^{-1} = \frac{1}{4}(1.8)\rho(0.7)\overline{V}_r = 1.27(\frac{1}{4}\rho\overline{V}_r)$).

7.2 Scattering of EM Radiation from Particles

As a final effect due to particulates, consider a cloud of particles that might form in the vicinity of a vehicle. Particulate contamination in the vicinity of spacecraft may interfere substantially with electromagnetic observations in the UV, visible, and IR regions of the spectrum. Particulates have been observed around manned spacecraft (Newkirk, 1967) including a number of *Shuttle* missions (Clifton and Owens, 1988) and most recently on the *Magellan* spacecraft (Goree and Chiu, 1993). On the *Magellan* spacecraft the optical scattering from the particulates was responsible for the periodic loss of the star-tracker lock. More generally, calculations show that remote observations will be affected by micrometer-class particles due to solar illumination, Earth radiation, and particulate self-emission (Rawlins and Green, 1987). If the particulates are moving slowly with respect to the spacecraft then they will pass slowly through the field of view and will provide a noise background that will be present much of the time. This may result in a malfunction of the instrument, which cannot be alleviated by simply pointing the instrument away from the Sun as was done with the *Magellan* mission. Hence, this may have a serious effect on mission operations.

There are several sources of particulates on orbit. First, particulates that collect on vehicle surfaces on the ground may be carried to orbit and shaken loose during spacecraft operations. Second, on a micrometer scale, many surfaces can be quite rough. Many composite materials may have fiber fragments that are micrometer size at the surface. Alternatively, under atomic oxygen attack, surfaces may become textured and flake on the micrometer scale. This was demonstrated on *LDEF*, which was flaking so much when brought into the *Shuttle* bay that it saturated the quartz crystal microbalances. In orbit, micrometer-size particulates can break off because of forces associated with solar illumination, thermal stresses due to terminator crossings, or electrostatic repulsion due to differential charging in the space plasma. Third, particulates are created by firing of rocket motors, especially solid rocket motors, and effluent dumps. It is well known that, for the *Shuttle*, water dumps can produce spectacular clouds of ice crystals. On vehicles where thrusters are placed without regard to plume impingement problems, surfaces where the plume impinges may be coated with large numbers of particulates or volatile exhaust products.

In addition to the natural sources of particulates that are discussed above, Goree and Chiu (1993) showed that exposure of a dusty spacecraft to the ionospheric plasma will result in enhanced dust shedding. This is because the dust particles become charged and are repelled from the spacecraft surface. Hence an analysis of the rate at which particulates are created must take into account the synergism with the plasma and neutral gas in the environment – an interesting and complex problem.

8

The State of the Art

8.1 Review of the Intellectual State of the Art

In this book, the concepts and theories behind spacecraft–environment interactions were examined. Each chapter is illustrated with selected examples. On the basis of these examples and theories, a reader should be able to identify and construct simple, first-order estimates of the principal interactions of importance to a specific spacecraft. The discipline of spacecraft–environment interactions represented by this process is, however, continuing to evolve both in its intellectual underpinnings and in its value to the spacecraft designer. In this final chapter, the current state of the art in spacecraft–environment interactions is reviewed and the future direction of progress in the field is predicted.

8.1.1 Neutrals

The primary interaction concerns for the neutral atmosphere are drag, atomic oxygen erosion, glow, and contamination. Although the overall processes associated with spacecraft drag in LEO or polar orbits are reasonably well understood, the detailed effects are often hard to predict accurately. The neutral environment for the Earth and its reaction to the Sun are, in principle, moderately well understood. Indeed, statistical models have been built that offer reasonable accuracy. However, there are still outstanding questions to be answered as to how solar and geophysical activity couple to the atmosphere and how to model the often almost impulsive atmospheric responses to sudden changes in these parameters. Errors as high as factors of 10 to 100 in predicting the density along an orbit are not unusual. Once the neutrals strike the spacecraft surface, the accommodation of the neutrals on the surface becomes an issue. Unfortunately, accommodation coefficients are still not well understood for many standard spacecraft surface materials. Accurate models of drag could be constructed if these accommodation coefficients for the interaction of atmospheric molecules with common spacecraft surfaces were better characterized.

As in the case of atmospheric drag, the generic atomic oxygen degradation process in LEO is known, but the details are not. To date, most studies of erosion rates of materials have been empirical. Although many different materials have been characterized in space experiments and erosion rates have been determined, there is still only a relatively poor understanding of the exact mechanism for the degradation. Therefore, the ability to predict how a new material will erode in space and hence the ability to develop tailored surface materials does not yet exist.

Because of its potentially deleterious effects on optical sensors, many measurements of spacecraft glow have been made from the *Shuttle*. Although a host of mechanisms have been postulated, it is now widely believed that one of the reactions between nitrous oxide and the oxygen impacting the surface is responsible for the glow. Indeed, up until very recently (July 1994), it was felt that this problem was well understood, but recent tests of the nitrous oxide process have yielded results opposite to what was predicted. Thus even this "understood" area is uncertain.

Spacecraft contamination by neutrals is an important driver in choosing materials because of outgassing and for placement of emission sources such as thrusters or plasma emitters. Thus it is not surprising that it is fairly well understood or at least reliably modeled. Line-of-sight contamination and reflection from surfaces are easy to calculate if the appropriate accommodation coefficients are understood. The backflow from thrusters and other effluent sources, although not perfectly understood, can also be calculated within less than an order of magnitude. Large-scale Monte Carlo codes and advanced computers have enabled the routine calculation of this backflow environment.

8.1.2 *Plasmas*

The primary concerns for plasma interactions with spacecraft at GEO are spacecraft surface charging and the subsequent arcing. The phenomenon of spacecraft charging is well understood and can be modeled with comparative accuracy provided the surface material properties are known. In contrast, the subsequent arcing is very poorly understood, although its effects can be well modeled using circuit analysis models. First, there are many possible mechanisms that can lead to arcing and the interaction between these, the plasma and the radiation from the environment, is not well modeled. Second, the actual initiation of an arc is not understood even when the process may be – mechanical stresses, EMP, cosmic-ray ionization, exceeding a threshold potential, and so forth may initiate the arc breakdown.

The use of plasma sources or thrusters on either GEO or LEO spacecraft is still a field for exploration. It is known that plasma sources can envelope a spacecraft in plasma and thereby discharge differential potentials on the spacecraft surfaces.

The dynamics of how this is done, the questions of where the plasma goes and how it responds to magnetic fields are all unanswered. Even the fundamental questions of the current patterns that the plasma allows and where the current paths close are only vaguely understood.

For spacecraft in LEO with unbiased surfaces, the level of the interactions of the plasma with the spacecraft is small and benign. For spacecraft with exposed high-voltage surfaces, there are a large number of unanswered questions. Current collection by such surfaces, especially in the presence of a neutral gas, their arcing, and the structure of the plasma flow around them are not well understood. The advent of electrodynamic tethers has raised new questions as to current closure around objects that are hundreds of kilometers long as well as to the impedance of the ionosphere for very large current densities. In fact, at the current densities required for significant power generation, the response of the ionosphere is almost certainly nonlinear. In addition to these issues, the question of anomalous ionization of emitted neutral gas from a spacecraft is still unanswered.

8.1.3 Radiation

The generic effects of radiation on materials are well understood. Radiation in space is a major driver in the choice of spacecraft electronics, materials, and solar cells because lack of attention to radiation damage will doom a mission. This area has benefitted enormously from advances in the companion field of nuclear physics. As a result, it is possible to predict the number of likely SEUs to be experienced by an IC before it flies or the radiation damage to a given material. However, our understanding of the radiation environment is much worse. The fundamental question of how the radiation belts are populated is still not adequately answered. Over the years, space experiments such as *CRRES* have shown that new radiation belts can appear following intense solar activity, last for a few months, and then apparently disappear. This inability to accurately predict the instantaneous or future radiation environment leads to overly conservative radiation design margins being imposed in the selection of spacecraft parts or shielding. This overdesign could be eliminated if the radiation sources were better understood. Internal charging of materials by high energy electrons needs also to be better understood.

8.1.4 Particulates

The fact that a hypervelocity impact with even a small particle can damage a spacecraft is well appreciated – a fleck of paint damaged a *Shuttle* window. The physics of hypervelocity impact has been extensively studied and can be modeled for simple cases. It is understood that particulates below about 1 cm in diameter can be shielded against, but larger particles should be avoided. The most serious gap in understanding comes in determining the actual population of debris and meteoroids to be found in the space environment. Current models are based on observational data, which are limited because of the inability to measure particulates in the 1- to 10-cm range. This uncertainty has the potential to seriously impact the design of the space station, which must protect humans and their systems for 10 to 15 years, by forcing a heavy shielding system on the designers. Finally, test facilities for velocities above $\sim$7 km/s are inadequate at present.

8.2 Review of the State of Current Engineering Practices

Current engineering practice in spacecraft–environment interactions is to design on the basis of experience, with only limited use of modeling codes (typically only radiation and thermal codes). Subsystem and system tests are performed on a spacecraft in ground-based facilities to validate the predictions – usually only thermal vacuum, vibro-acoustics, and EMC/EMP. Radiation testing is normally only carried out on individual piece parts. Modeling of spacecraft charging, meteoroid and/or space debris impacts, contamination, or oxygen erosion are typically only considered if specifically called out in the requirements or if the systems are expected to be particularly sensitive. Although this procedure may be successful for one-of-a-kind spacecraft in LEO, long-lasting or multiple spacecraft systems could be seriously underdesigned. In particular, synergism might produce unexpected effects that could be missed by these current design practices.

8.2.1 Interaction Modeling Codes

The state of the art in developing tools for modeling spacecraft interactions as a whole is in good shape. Simulation tools have been developed to help designers in modeling the first-order effects of spacecraft–environment interactions on a spacecraft. These are the Environment Work Bench (EWB) and ENVIRONET in the United States and ESA Base in Europe. These tools incorporate simple models of the effects discussed in this book. The tools are designed to be accessible from, or run on, typical computers and can be used to evaluate different configurations, choices of materials, orbits, and so forth.

For detailed calculations of neutral effects, there are two- and three-dimensional Monte Carlo codes that simulate backflow and atomic oxygen erosion. For contamination and plume structures, there are the CONTAM II and SOCRATES codes. In the plasma interactions, codes such as NASCAP, NASCAP/LEO, and POLAR are available. In the radiation interactions area, in addition to the AE8/AP8 trapped-radiation environment codes, there are several radiation shielding codes such as

NOVICE and SHIELDOSE. Finally, NASA has made available a series of models describing the micrometeoroid and debris environments and their effects that yield useful, if conservative, estimates of the expected penetration probabilities as a function of particle size.

8.2.2 Ground-Based Experimental Capabilities

In addition to radiation sources and charged-particle accelerators, a common ground-based environmental spacecraft test facility will contain a large vacuum chamber where a spacecraft can be exposed to the low pressures of space under simulated solar and/or thermal conditions. There are usually companion facilities for conducting EMI/EMC tests between different parts of a spacecraft as well as vibro-acoustic tests. The past few years have even seen the creation of facilities to simulate atomic oxygen erosion. These produce a supersonic atomic oxygen beam that then bombards selected spacecraft surface materials. There are large plasma tanks to examine the influence of high-voltage surfaces in space. Hypervelocity impact facilities exist in the United States and abroad, capable of simulating velocities up to 10 to 15 km/s, though the size and range of particles are severely limited. Thus many of the effects of importance to spacecraft interactions can be simulated.

8.3 Future Trends in the Practice of
Spacecraft–Environment Interactions

The preceding has provided a short synopsis of the current state of the art in spacecraft interactions. From this state of affairs, where will environmental interactions be a decade from now? Given the increasing uncertainty in the space program, it is difficult to predict where the industry will be in five years, let alone one or two decades. There are, however, several approaches being taken in the area of spacecraft design that involve specific guidelines for how to address systems' reliability, and a great deal can be said about the likely future of requirements for modeling the environment and its effects. These approaches to spacecraft design can be loosely grouped in terms of their mission classification. Specifically, within the NASA community there has developed a classification system based on the desired reliability of a mission. Table 8.1 provides a definition of four classes of missions.

The details of the definitions in Table 8.1 vary from institution to institution, but the implication is clear: for Class A and B missions, typically no expense is spared in defining the environment. The interactions are modeled in detail and supplemented by actual experiments and tests where possible. Class C missions follow a fairly well-defined procedure, depending on the institution – models are normally well defined internally and with test procedures and facilities in place. Interestingly, it

Table 8.1. *Definitions of spacecraft classes*

Class	Definition
A	World-class mission. National prestige, defense, or human life on the line. Typically expensive ($> \$1B$), long-duration, interplanetary mission like *Galileo*. Ultra-high reliability, multiple redundant systems. Nonrepairable.
B	Important mission. National prestige important. May be repairable (from *Shuttle*). Expensive, long duration. Very-high reliability, no single string failures.
C	Typical mission. 3- to 5-year lifetime. Single string failures acceptable if risk evaluated. "Standard" reliability program.
D	Fly it if it makes the launch site on time and can't hurt other experiments or spacecraft on the same flight.

is the Class D type of mission, where cost or schedule are likely paramount, that space environment engineering comes into its own. With the desire for large fleets of many similar (and cheap) spacecraft, there is a fine line between just enough reliability and too much or too little. As an example of these issues, the cost of the overall reliability program (and proportionally, environmental engineering effort) can run as high as 30 percent of the spacecraft cost for a Class A or B mission. This percentage drops to $\sim$10 percent for a typical Class C mission and may go to zero for a Class D. Yet, with the need to build many cheap spacecraft (e.g., for one of the proposed LEO communications systems), small changes in design to accommodate environmental concerns (such as varying the radiation shielding weight) can drive large changes in cost or reliability. Thus it is the Class D mission that will likely be the greatest challenge in the future as we struggle to determine the proper proportion of environmental engineering to apply.

8.3.1 High-Reliability Missions

Although Class D missions may present the greatest challenge for the space environment engineer, there are several important challenges for the Class A and B missions. First, there is the continual need to update existing models, and second, as new environments and new interactions are encountered, there is the exciting opportunity to develop new models and procedures. As discussed, even a few short years ago, there was little appreciation for space debris or oxygen erosion at LEO – perhaps the best known of the space environment regions. Indeed, as systems become more complex, we expect the interactions that affect them to become more subtle and complex – witness the growing issues associated with SEEs which are no longer limited just to SEUs, but include latchup, burnout, gate rupture, and so forth. Similarly, the environments that we will encounter also will be more complex. As an excellent example of this, consider either a base on the lunar surface or an O'Neil colony. Each of these systems will interact with the space environment in unique and potentially devastating ways. In the case of the lunar base, simply traversing the lunar surface will kick up dust and debris – this environment will affect ultrasensitive optical and particle detectors on the lunar surface. O'Neil colonies will leak air at a prodigious rate – through broken windows, leaky airlocks, control thruster firings, docking operations, and so forth. After a time, the cloud of gaseous products around the colony (or, on a smaller scale, the space station) will be substantial (leading to glow phenomena, surface interactions, and contamination) and may even begin to effect the Earth's magnetosphere by deviating the solar wind. Such issues, rarely considered in the past, will drive the environmental engineering for the Class A and B programs of future generations.

8.3.2 Class C Programs

The major thrust in the area of Class C programs is the increasing number of attempts to link all of the different types of environmental and interactions models together. With products such as ESA Base (ESA), ENVIRONET (NASA GSFC), or EWB/EPSAT (DoD/NASA), government organizations and private industry are making great strides toward the 'universal spacecraft design tool'. This exciting development means that, when the problems are worked out with these techniques, there will be a computer design tool like NASTRAN for the aeronautics world that can be used to design in the environmental reliability of a spacecraft from the beginning – the engineer puts in the requirements and the desired range, the program recommends an orbit, a basic design, and a test specification. Ultimately, this tool will allow the environmental engineer to test different packages of requirements against varying designs and cost constraints, allowing a wide range of options for the program or project office. Although such a universal tool will not answer every type of issue and will probably not have the accuracy or fidelity required for a Class A or Class B mission, it will permit many smaller projects and organizations to develop quality products efficiently.

8.3.3 Class D Programs

The Class D program may offer the most interesting challenges for the environmental engineer. The reason for this claim is that the tremendous temporal variations in the environmental parameters that we have seen in preceding chapters will force careful consideration of the environmental risk for each phase of a mission if a reasonable chance for success is expected. It means that a design that is to incorporate exotic technologies and be done at low costs will require a good understanding of the actual risk up front. Risk due to the environment, its definition and assessment, will be at the heart of the success of many Class D missions. Given the uncertainty in the environment and its effects over a given time period, risk assessment will be a necessary component of any design effort or mission plan.

Protection from the effects of the environment is potentially costly – both in terms of the protection methods (extra weight, power, or even computing power) and the time spent in design changes or in testing. This protection can be the anathema of the "faster, cheaper, better" type of Class D mission. (*Note:* Not all such missions are Class D, nor are Class D missions necessarily this type – they are closely related, however, given the current definition of "faster, cheaper, better" being used throughout the industry.) Testing against environmental constraints can be particularly costly in terms of schedule and added stress on the spacecraft. Often, in this class of program, the vehicle or system is the first of a kind – no engineering test model exists. Overtesting the system can dramatically reduce its on-orbit life. Likewise, undertesting or too low an environmental requirement or design can doom the mission. Thus a fine line between little concern for environmental effects and too much needs to be drawn.

What will be the role of the environmental engineer in this type of mission in the future? Clearly, the first step is to be able to rapidly assess the minimum environmental requirements likely to be applicable to the mission. This text has been aimed at developing a rational procedure for accomplishing that determination. The second step must be to properly establish the risks associated with the primary environmental drivers. In the case of radiation effects, typically a primary driver for most missions, this can be reduced to determining the likelihood of a solar proton event or of encountering a certain level of trapped radiation. Currently, the models for assessing these effects are undergoing revision – an observation likely to still be true a decade or more from now. The future will hopefully see both a growth in our environmental prediction techniques and in definition of the details of the interactions. Already this is happening in the microelectronics industry. For example, with the growing sensitivity of microelectronics subsystems such as solid-state data recorders, design engineers have worked out built-in error detection and correction algorithms for limiting the effects of SEUs. In the future, all segments of the environmental engineering community will need to keep abreast of such fixes and under what conditions they apply. This will be particularly true of the cutting-edge, high-risk Class D mission.

As in the case of the Class C spacecraft, the universal environmental tool may turn out to be a valuable means of keeping down costs and allowing trade-offs early on in the project. Unfortunately, often by their very nature (i.e., rapid turnaround or high-risk technologies), Class D missions may not be able to implement all of the standard design fixes. As an example, the recent *Clementine* mission saw the first large-scale use of plastic-encapsulated devices (PEDs) or plastic parts as they are called. The tremendous and rapid growth in the computer world has meant that even the lowliest home computer can have many times the capability of the costliest, ultrareliable spacecraft computer. Conventional wisdom indicates that the environmental effects on PEDs are so overwhelming that they cannot successfully be flown in space. Yet such devices are now being used in a variety of extremely stressful environments here on Earth, such as inside automobile engines. Even the minimal reliability requirements of a Class C mission would not be met by such parts, yet they are clearly much more capable and cheaper than any existing space technology. This is where the environmental engineer comes in – on *Clementine*, after careful consideration of all aspects of the parts reliability process, PEDs were tested and incorporated throughout the spacecraft design. Some of the parts had radiation sensitivities of less than 1 krad (Si), yet with proper consideration of the environment (the mission did not spend much time in the radiation belts and it flew at Solar Minimum), it became clear that such a mission would be feasible. Even so, there would have been significant radiation problems if the spacecraft had encountered a 1972 level solar proton event. The program office, after careful evaluation of all aspects of the risk, determined that the gains in our understanding of how PEDs would work in space versus the probability of a proton event were well worth the risk. (*Note:* A proton event was observed during the *Clementine* mission but it was well under the conservative levels adopted by the program, and no effects on spacecraft operations were observed.) Finally, even the definitions A, B, C, and D are being phased out in response to mission-tailored risk definitions. Even so, the concepts raised should be valuable in both assessing and controlling risk due to environmental interactions.

8.3.4 Summary

It is impossible in this period of exponential growth in technology to accurately predict where new environmental concerns will arise. It is certain, however, that as long as risk plays an important role in mission design, informed, intelligent consideration of the effects of the space environment will be a significant component of a spacecraft's design. Given the dramatically different goals of individual space missions, it is little wonder that there are many levels of environmental risk assessment. Thus the accurate assessment of risk will drive the future of space-environment engineering and will define the field. It is ultimately toward this end that the lessons in this text are directed.

References

Adamo, R. C., and Matarrese, J. R. Transient pulse monitor data from the P78-2 (SCATHA) spacecraft, *Journal of Spacecraft and Rockets*, 20:432–7, 1983.

Adams, J. H. The Ionizing Particle Environment near Earth, Technical Report AIAA 82-107, American Institute of Aeronautics and Astronautics, Washington, DC, 1982.

Adams, J. H., Letaw, J. R., and Smart, D. F. Cosmic Ray Effects on Microelectronics, Part II: The Geomagnetic Cutoff Effects. NRL Memorandum Report 4506, Naval Research Laboratory, 1983.

Adams, J.H. Cosmic Ray Effects on Microelectronics, Part IV: NRL Memorandum Report 5901, Naval Research Laboratory, 1986.

Agrawal, B. N. *Design of Geosynchronous Spacecraft*, Prentice-Hall, Inc., Englewood Cliffs, NJ, 1986.

Al'pert, Ya. L. *The Near-Earth and Interplanetary Plasma*, Vols. 1 and 2, Cambridge University Press, 1983.

Al'pert, Ya. L., Gurevich, A. V., and Pitaevskii, L. P. *Space Physics with Artificial Satellites*. Consultants Bureau, 1965.

Balmain, K. G. Surface discharge effects, In H. B. Garrett and C. P. Pike (eds.), *Space Systems and Their Interactions with the Earth's Space Environment*, Vol. 71, pp. 276–98, American Institute of Aeronautics and Astronautics, Washington, DC, 1980.

Balsiger, H., Eberhardt, P., Geiss, J., and Young, D. T. Magnetic storm injection of 0.9- to 16-keV/e solar and terrestrial ions into the high-altitude magnetosphere, *Journal of Geophysical Research*, 85:1645–62, 1980.

Banks, B., Rutledge, S. K., Auer, B., and DiFilipo, F. Atomic oxygen undercutting of defects on SiO_2 protected polyimide solar array blankets, In *Materials Degradation in Low Earth Orbit (LEO)*. Minerals, Metals, and Materials Society, 1990.

Banks, P. M., and Kockarts, G. *Aeronomy, Parts A and B*, Academic Press, New York, 1973.

Banks, P. M., Williamson, P. R., and Raitt, W. J. Space shuttle glow observations, *Geophysical Research Letters*, 10:118, 1983.

Bareiss, L. E., Payton, R. M., and Papaziaa, H. A. Shuttle/Spacelab Contamination Environment and Effects Handbook – Second Edition, Martin Marietta Aerospace, MCR-85-583, 1986.

Barnett, A., and Olbert, S. Radiation of plasma waves by a conducting body moving through a magnetized plasma, *Journal of Geophysical Research*, 91:10117–35, 1986.

Barraclough, D. R. International geomagnetic reference field: the 4th generation, *Physics of the Earth and Planetary Interiors*, 48:279, 1987.

Belcastro, V., Veltri, P., and Dobrownoly, M. Radiation from long conducting tethers moving in the near-earth environment, *Nuovo Cimento*, 5:537–60, 1982.

Bernstein, I. B., and Rabinowitz, I. N. Theory of electrostatic probes in a low-density plasma, *Physics of Fluids*, 2:112, 1959.

Besse, A. L., and Rubin, A. G. A simple analysis of spacecraft charging involving blocked photoelectron currents, *Journal of Geophysical Research*, 85(A5):2324–8, 1980.

Bird, G. A. *Molecular Gas Dynamics*, Clarendon Press, Oxford, England, 1976.

Bittencourt, J. A. *Fundamentals of Plasma Physics*, Pergamon Press, New York, 1986.

Bourrieau, J. Protection and shielding, In R. N. DeWitt, D. Duston, and A. K. Hyder (eds.), *The Behaviour of Systems in the Space Environment*, pp. 299–351, Kluwer Academic Publishers, 1993.

Carruth, M. R., and Brady, M. E. Measurement of the charge-exchange plasma flow from an ion thruster, *Journal of Spacecraft and Rockets*, 18(5):457–61, 1981.

Chaky, R. C., Nonnast, J. H., and Enoch, J. Numerical simulation of the sheath structure and current–voltage characteristics of a conductor-dielectric disk in a plasma, *Journal of Applied Physics*, 52:7092–8, 1981.

Chen, F. F. *Introduction to Plasma Physics and Controlled Fusion*, Plenum Press, New York, 1984.

Cho, M., and Hastings, D. E. Dielectric charging processes and arcing rates of high-voltage solar arrays, *Journal of Spacecraft and Rockets*, 28(6):698–706, 1991.

Clifton, K. S., and Owens, J. K. Optical contamination measurements on early Shuttle missions, *Applied Optics*, 27:603, 1988.

Cour-Palais, B. G. Meteoroid environment model – 1969 (near Earth to lunar surface), Technical Report NASA SP-8013, NASA, 1969.

Cour-Palais, B. G. Hypervelocity impact in metals, glass, and composites, *International Journal of Impact Engineering*, 5:221–37, 1987.

Craven, P. D. Potential modulation on the SCATHA satellite, *Journal of Spacecraft and Rockets*, 24:150–7, 1987.

DeForest, S. E. Spacecraft charging at synchronous orbit, *Journal of Geophysical Research*, 77:3587–3611, 1972.

Dettleff, G. Plume flow and plume impingement in space technology, *Progress in Aerospace Science*, 28:1–71, 1991.

DeWitt, R. N., Duston, D., and Hyder, A. K. *The Behaviour of Systems in the Space Environment*, Kluwer Academic Publishers, 1993.

Divine, T. N., and Garrett, H. B. Charged-particle distributions in Jupiter's magnetosphere, *Journal of Geophysical Research*, 88:6889–6903, 1983.

Divine, T. N. Five populations of interplanetary meteoroids, *Journal of Geophysical Research*, 98(E9):17029–48, 1993.

Dobrownoly, M., and Melchioni, E. Expansion of polarized plasma source into an ambient plasma, *Journal of Plasma Physics*, 47:111, 1992.

Dobrownoly, M., and Melchioni, E. Electrodynamic aspects of the first tethered satellite mission, *Journal of Geophysical Research*, 98(A8):13761–78, 1993.

Donahue, D. J., Neubert, T., and Banks, P. M. Estimated radiated power from a conducting tethered satellite system, *Journal of Geophysical Research*, 96(A12):21245–53, 1991.

Feynman, J., Spitale, G., Wang, J., and Gabriel, S. Interplanetary proton fluence: JPL 1991, *Journal of Geophysical Research*, 98(A8):13281–94, 1993.

Fredrickson, A. R. Radiation-induced dielectric charging, In H. B. Garrett and C. P. Pike (eds.), *Space Systems and Their Interactions with the Earth's Space Environment*, Vol. 71, pp. 386–412, American Institute of Aeronautics and Astronautics, Washington, DC, 1980.

Fredrickson, A. R., Cotts, D. B., Wall, J. A., and Bouquet, F. L. *Spacecraft Dielectric Material Properties and Spacecraft Charging*, Vol. 107, of *AIAA Progress in Aeronautics and Astronautics*, American Institute of Aeronautics and Astronautics, Washington, DC, 1986.

Fredrickson, A. R., Holeman, E. G., and Mullen, E. C. Characteristics of spontaneous electrical discharging of various insulators in space radiations, *IEEE Transactions of Nuclear Science*, 39(5–6):1773–1782, December 1992.

Garrett, H. B. The charging of spacecraft surfaces, *Reviews of Geophysics*, 19:577–616, 1981.

Garrett H. B., and DeForest, S. E. An analytical simulation of the geosynchronous plasma environment, *Planetary Space Science*, 27:1101–09, 1979.

Garrett, H. B., and Pike, C. (eds.) *Space Systems and Their Interactions with Earth's Space Environment*, American Institute of Aeronautics and Astronautics, Washington, DC, 1980.

Garrett, H. B., Schwank, D. C., and DeForest, S. E. A statistical analysis of the low-energy geosynchronous plasma environment–I. electrons, *Planetary Space Science*, 29:1021–44, 1981a.

Garrett, H. B., Schwank, D. C., and DeForest, S. E. A statistical analysis of the low-energy geosynchronous plasma environment–II. protons, *Planetary Space Science*, 29:1045–60, 1981b.

Garrett, H. B., and Spitale, G. C. Magnetospheric plasma modeling (0–100 keV), *Journal of Spacecraft and Rockets*, 22:231–44, 1985.

Garrett, H. B., Chutjian, A., and Gabriel, S. Space vehicle glow and its impact on spacecraft systems, *Journal of Spacecraft and Rockets*, 25:321, 1988.

Garrett, H. B., and Petrasek, I. Probability of environmentally induced failure during Galileo VEEGA due to meteroid impact, radiation damage, or spacecraft changing, Jet Propulsion Laboratory Report JPL D-5832, The Jet Propulsion Laboratory, 1988.

Geiss, J., Balsiger, H., Eberhardt, P., Walker, H. P., Weber, C., Young, D. J., and Rosenbauer, H. Dynamics of magnetospheric ion composition as observed by the GEOS mass spectrometer, *Space Science Reviews*, 22:537, 1978.

Gerver, M. J., Hastings, D. E., and Oberhardt, M. Theory and experimental review of plasma contactors, *Journal of Spacecraft and Rockets*, 27:391–402, 1990.

Goldstein, R., and DeForest, S. E. Active control of spacecraft potentials at geosynchronous orbit, In *Spacecraft Charging by Magnetospheric Plasmas*, Vol. 47, pp. 169–81. American Institute of Aeronautics and Astronautics, Washington, DC, 1976.

Goller, G. R., and Grun, E. Calibration of the *Galileo/Ulysses* dust detectors with different projectile materials and at varying impact angles, *Planetary and Space Science*, 37:1197–1206, 1989.

Goree, J., and Chiu, Y. T. Dust contamination of the spacecraft environment by exposure to plasma, *Journal of Spacecraft and Rockets*, 30(6):765–7, 1993.

Grard, R. J. Properties of the satellite photoelectron sheath derived from photoemission laboratory measurements, *Journal of Geophysical Research*, 78:2885–2906, 1973.

Grard, R., Knott, K., and Pederson, A. Spacecraft charging effects, *Space Science Reviews*, 34:239–304, 1983.

Green, B. D., Caledonia, G. E., and Wilkerson, T. D. The shuttle environment: Gases, particulates and glow, *Journal of Spacecraft and Rockets*, 22:500–11, 1985.

Grier, N. T. Plasma interaction experiment II: laboratory and flight results, In *Spacecraft Environment Interactions Technology Conference*, pp. 333–48, NASA CP-2359, 1983.

Gull, T. R., Herzog, H., Osantowski, J. F., and Toft, A. R. Low earth orbit environmental effects on osmium and related optical thin film coatings, *Applied Optics*, 24(16):2660–5, 1985.

Gussenhoven, M. S., and Mullen, E. G. Geosynchronous environment for severe spacecraft charging, *Journal of Spacecraft and Rockets*, 20:26–34, 1983.

Gussenhoven, M. S., Hardy, D. A., Rich, F., Burke, W. J., and Yeh., H. C. High-level spacecraft charging in the low-altitude polar auroral environment, *Journal of Geophysical Research*, 90:11009–23, 1985.

Hardy, D. A., Gussenhoven, M. S., and Holeman, E. A statistical model of auroral electron precipitation, *Journal of Geophysical Research*, 90:4229–48, 1985.

Hardy, D. A., Gussenhoven, M. S., and Brautigam, D. A statistical model of auroral ion precipitation, *Journal of Geophysical Research*, 94:370–92, 1989.

Hastings, D. E. The use of electrostatic noise to control high-voltage differential charging of spacecraft, *Journal of Geophysical Research*, 91:5719–24, 1986.

Hastings, D. E. Enhanced current flow through a plasma cloud by induction of plasma turbulence, *Journal of Geophysical Research*, 92:7716–22, 1987a.

Hastings, D. E. Theory of plasma contactors used in the ionosphere, *Journal of Spacecraft and Rockets*, 24:250–6, 1987b.

Hastings, D. E., and Wang, J. The radiation impedance of a electrodynamic tether with end connectors, *Geophysical Research Letters*, 14:519–22, 1987.

Hastings, D. E., and Gatsonis, N. A. Plasma contactors for use with electrodynamic tethers for power generation, *Acta Astronautica*, 17:827–36, 1988.

Hastings, D. E., and Blandino, J. Bounds on current collection by plasma clouds from the ionosphere, *Journal of Geophysical Research*, 94:2737–44, 1989.

Hastings, D. E., and Chang, P. The physics of positively biased conductors surrounded by dielectrics in contact with a plasma, *Physics of Fluids B*, 1:1123–32, 1989.

Hastings, D. E., and Cho, M. Ion drag for a negatively biased solar array in LEO, *Journal of Spacecraft and Rockets*, 27:279–84, 1990.

Hastings, D. E., Weyl, G., and Kaufman, D. A simple model for the threshold voltage for arcing on negatively biased high-voltage solar arrays, *Journal of Spacecraft and Rockets*, 27:539–44, 1990.

Hastings, D. E., Cho, M., and Kuninaka, H. The arcing rate for a high voltage solar array: theory, experiments, and predictions, *Journal of Spacecraft and Rockets*, 29(4):538–54, 1992.

Hastings, D. E., Cho, M., and Wang, J. The Space Station Freedom structure floating potential and the probability of arcing, *Journal of Spacecraft and Rockets*, 29(6):830–4, 1992.

Hedin, A. E. A revised thermospheric model based on mass spectrometer and incoherent scatter data: MSIS-83, *Journal of Geophysical Research*, 88:10170–88, 1983.

Hedin, A. E. MSIS-86 thermospheric model, *Journal of Geophysical Research*, 92:4649–62, 1987.

Hedin, A. E. Extension of the MSIS thermospheric model into the middle and lower atmosphere, *Journal of Geophysical Research*, 96:1159–72, 1991.

Hoffman, A. Galileo orbiter functional requirements book, environmental design requirements, Technical Report GLL-3-240, Rev. C, Jet Propulsion Laboratory, 1987.

Holmes-Siedle, A., and Adams, L. *Handbook of Radiation Effects*, Oxford University Press, Oxford, England, 1993.

Humes, D. H., Alvarez, J. M., O'Neal, R. L., and Kinnard, W. H. The interplanetary and near-Jupiter meteoroid environment, *Journal of Geophysical Research*, 79:3677–84, 1974.

Hurlbut, F. C. *Gas/Surface Scatter Models for Satellite Applications*, Vol. 103 of *Thermophysical Aspects of Reentry Flows*, pp. 97–119, American Institute of Aeronautics and Astronautics, Washington, DC, 1986.

Iess, L., and Dobrownoly, M. The interaction of a hollow cathode with the ionosphere, *Physics of Fluids*, B1:1880–9, 1989.

Jacobs, J. A. *Geomagnetic Micropulsations*, Springer–Verlag, New York, 1970.

Jemiola, J. M. Spacecraft contamination: A review, In H. B. Garrett and C. P. Pike (eds.), *Space Systems and Their Interactions with the Earth's Space Environment*, Vol. 71, pp. 680–706. American Institute of Aeronautics and Astronautics, Washington, DC, 1980.

Johnson, N. L., and McKnight, D. S. *Artifical Space Debris*, Krieger Publishing Co., Melbourne, FL, 1991.

Jongeward, G., Katz, I., Mandell, M., and Parks, D. E. The role of unneutralized surface ions in negative potential arcing, *IEEE Transactions on Nuclear Science*, NS-32(6):4087–91, 1985.

Jordan, T. M. NOVICE: A Radiation Transport/Shielding Code: Users' Guide, Technical Report 87.01.02.01, Experimental and Mathematical Physics Consultants, 1987.

Jursa, A. S. (ed.) *Handbook of Geophysics and the Space Environment*, Air Force Geophysics Lab, 1985, NTIS Accession No. AD-A167000.

Kasha, M. A. *The Ionosphere and Its Interaction with Satellites*, Gordon and Breach, New York, 1969.

Katz, I., Parks, D. E., Mandell, M. J., Harvey, J. M., Wang, S. S., and Roche, J. C. NASCAP, a three-dimensional charging analyzer program for complex spacecraft, *IEEE Transactions on Nuclear Science*, NS-24(6):2276, 1977.

Katz, I., Mandell, M. J., Jongeward, G. A., and Gussenhoven, M. S. The importance of accurate secondary electron yields in modeling spacecraft charging, *Journal of Geophysical Research*, 91:13739–44, 1986.

Katz, I., Jongeward, G. A., Davis, V. A., Mandell, M. J., Kuharski, R. A., Lilley, J. R., Raitt, W. J., Cooke, D. L., Torbert, R. B., Larson, G., and Rau, D. Structure of the bipolar plasma sheath generated by SPEAR 1, *Journal of Geophysical Research*, 94:1450–8, 1989.

Kelley, M. C. *The Earth's Ionosphere: Plasma Physics and Electrodynamics*, Academic Press, San Diego, CA, 1989.

Kessler, D. J. Meteoroids and Orbital Debris, Technical Report SSP 30425, NASA, 1991.

Kessler, D. J. Orbital debris environment in low earth orbit: an update, *Advances in Space Research*, 13(8):139–48, 1993.

Kessler, D. J., and Cour-Palais, B. G. Collision frequency of artificial satellites: creation of a debris belt, In H. B. Garrett and C. P. Pike (eds.), *Space Systems and Their Interactions with the Earth's Space Environment*, Vol. 71, pp. 707–36. American Institute of Aeronautics and Astronautics, Washington, DC, 1980.

Kogan, M. N. *Rarefied Gas Dynamics*, Plenun Press, New York, 1969.

Koons, H. C., Mizera, P. F., Roeder, J. L., and Fennell, J. F. Severe spacecraft charging event on SCATHA in September 1982, *Journal of Spacecraft and Rockets*, 25:239–43, 1988.

Koons, H. C., and Gorney, D. J. Relationship between electrostatic discharges on spacecraft P78-2 and the electron environment, *Journal of Spacecraft and Rockets*, 28:683–8, 1991.

Koons, H. C. Summary of environmentally induced electrical discharges on the P78-2 (SCATHA) satellite, *Journal of Spacecraft and Rockets*, 20:425–31, 1983.

Koontz, S. L., Albyn, K., and Leger, L. J. Atomic oxygen testing with thermal atom systems: A critical evaluation, *Journal of Spacecraft and Rockets*, 28(3):315, 1991.

Krall, N. A., and Trivelpiece, A. W. *Principles of Plasma Physics*, McGraw–Hill, New York, 1973.

Krech, R. H., Gauthier, M. J., and Caledonia, G. E. High velocity atomic oxygen/surface accomodation studies, *Journal of Spacecraft and Rockets*, 30:509–13, 1993.

Laframboise, J. G., and Sonmor, L. J. Current collection by probes and electrodes in space magnetoplasmas: a review, *Journal of Geophysical Research*, 98(A1):337–57, 1993.

Lai, S.T. Theory and observation of triple-root jump in spacecraft charging, *Journal of Geophysical Research*, 96(A11):19269–81, 1991a.

Lai, S.T. Spacecraft charging thresholds in single and double Maxwellian space environments, *IEEE Transactions on Nuclear Science*, 38(6):1629–33, 1991b.

Lam, S., and Greenblatt, M. On the interaction of a solid body with a flowing collisionless plasma, In *The Fourth Symposium on Rarefied Gas Dynamics*, Academic Press, 1966.

Latham, R. V. Potential threats to the performance of vacuum-insulated high-voltage devices in a space environment, In R. N. DeWitt, D. Duston, and A. K. Hyder (eds.), *The Behaviour of Systems in the Space Environment*, pp. 467–90. Kluwer Academic Publishers, 1993.

Laurance, M. R., and Brownlee, D. E. The flux of meteoriods and orbital space debris striking satellites in low earth orbit, *Nature*, 323:136–38, 1986.

Leger, L. L., and Visentine, J. T. A consideration of atomic oxygen interactions with the space station, *Journal of Spacecraft and Rockets*, 23(5):505–11, 1986.

Leung, P., Whittlesey, A. C., Garrett, H. B., Robinson, P. A., and Divine, T. N. Environment-induced electrostatic discharges as the cause of *Voyager 1* power-on resets, *Journal of Spacecraft and Rockets*, 23:323–30, 1986.

Linson, L. M. Current–voltage characteristics of an electron-emitting satellite in the ionosphere, *Journal of Geophysical Research*, 74:2368, 1969.

Lucas, A. A. Fundamental processes in particle and photon interactions with surfaces, In R. J. L. Grard (ed.), *Photon and Particle Interactions with Surfaces in Space*, pp. 3–21, Reidel, 1973.

Maag, C. R. Results of apparent atomic oxygen reactions with spacecraft materials during Shuttle flight STS-41g, *Journal of Spacecraft and Rockets*, 25(2):162, 1988.

Martin, D. J., and Maag, C. R. The influence of commonly used materials and compounds on spacecraft contamination, In *IAF 92-0336*, International Astronautical Federation, 1992.

Martinez-Sanchez, M., and Hastings, D. E. A systems study of a 100-kW tether, *Journal of Astronautical Sciences*, 35:75–96, 1987.

Mullen, E., and Gussenhoven, M. Results of space experiments: CRRES, In R. N. DeWitt, D. Duston, and A. K. Hyder (eds.), *The Behaviour of Systems in the Space Environment*, pp. 605–54, Kluwer Academic Publishers, 1993.

Mullen, E. G., Gussenhoven, M. S., and Hardy, D. A. SCATHA survey of high-voltage spacecraft charging in sunlight, *Journal of Geophysical Research*, 91:1474–90, 1986.

Nanevicz, J. E., and Adamo, R. C. Occurrence of arcing and its effects on space systems, In H. B. Garrett and C. P. Pike (eds.), *Space Systems and Their Interactions with the*

Earth's Space Environment, Vol. 71, pp. 252–75, American Institute of Aeronautics and Astronautics, Washington, DC, 1980.

Neubert, T., and Banks, P. M. Recent results from studies of electron beam phenomena in space plasmas, *Planetary Space Science*, 40(2/3):153–83, 1992.

Neubert, T., Banks, P. M., Gilchrist, B. E., Fraser-Smith, A. C., Williamson, P. R., Raitt, W. J., Myers, N. B., and Sasaki, S. The interaction of an artificial electron beam with the earth's upper atmosphere: effects on spacecraft charging and the near plasma environment, *Journal of Geophysical Research*, 95:12209, 1990.

Newell, P. T. Review of the critical ionization velocity effect in space, *Review of Geophysics Research*, 23:93–104, 1985.

Newkirk, G. The optical environment of manned spacecraft, *Planetary Space Science*, 15:1267, 1967.

Nicolis, G. and Prigogine, I. *Self-Organization in Nonequilibrium Systems*, John Wiley, New York, 1977.

Papadopoulos, K. Scaling of the beam-plasma discharge for low magnetic fields, *Journal of Geophysical Research*, 91:1627, 1986.

Parker, L. W. and Murphy, B. L. Potential buildup on an electron-emitting ionspheric satellite, *Journal of Geophysical Research*, 72:1631, 1967.

Parks, D., and Katz, I. Theory of plasma contactors for electrodynamic tethered satellite systems, *Journal of Spacecraft and Rockets*, 24:245–9, 1987.

Parks, D., Mandell, M. J., and Katz, I. Fluid model of plasma outside a hollow cathode neutralizer, *Journal of Spacecraft and Rockets*, 19:354–7, 1982.

Particle Data Group, Passage of particles through matter; review of particle properties, *Physics Letters B*, 239(111):1–38, 1990.

Peters, P. N., Linton, R. C., and Miller, E. R. Results of apparent atomic oxygen reactions on Ag, C, and Os exposed during Shuttle STS-4 orbits, *Geophysical Research Letters*, 10:569–71, 1983.

Peters, P. N., Gregory, J. C., and Swann, J. T. Effects on optical systems from interactions with oxygen atoms in low earth orbits, *Applied Optics*, 25(8):1290–8, 1986.

Prokopenko, S. M. L., and Laframboise, J. G. High-voltage differential charging of geostationary spacecraft, *Journal of Geophysical Research*, 85:4125, 1980.

Purvis, C. K., and Bartlett, R. O. Active control of spacecraft charging, In H. B. Garrett and C. P. Pike (eds.), *Space Systems and Their Interactions with the Earth's Space Environment*, Vol. 71, pp. 299–317, American Institute of Aeronautics and Astronautics, Washington, DC, 1980.

Purvis, C., Garrett, H., Whittlesey, A. C., and Stevens, N. J. Design guidelines for assessing and controlling spacecraft charging effects, Technical Paper 2361, NASA, 1984.

Rault, D. F. G., and Woronowicz, M. S. Spacecraft contamination investigation by direct simulation Monte Carlo–contamination on UARS/HALOE, In *AIAA 93-0724*. American Institute of Aeronautics and Astronautics, Washington, DC, 1993.

Rawer, K. International reference ionosphere, *Advances in Space Research*, 2:181–257, 1982.

Rawlins, W. T., and Green, B. D. Spectral signatures of micron sized particles in the Shuttle optical environment, *Applied Optics*, 26:3052, 1987.

Robinson, Paul A., Jr. (ed.), Introduction to Spacecraft environments and the anomalies they cause, Jet Propulsion Laboratory, D-5489, California Insitute of Technology, Pasadena, CA, 1988.

Rosen, A. *Spacecraft Charging by Magnetospheric Plasmas*, American Institute of Aeronautics and Astronautics, Washington, DC, 1976.

Rubin, A. L., Katz, I., Mandell, M., Schnuelle, G., Steen, P., Parks, D., Cassidy, J., and Roche, J. A three-dimensional spacecraft-charging computer code, In H. B. Garrett and C. P. Pike (eds.), *Space Systems and Their Interactions with the Earth's Space Environment*, Vol. 71, pp. 318–36. American Institute of Aeronautics and Astronautics, Washington, DC, 1980.

Sagalyn, R. C., and Bowhill, S. A. Progress in geomagnetic storm prediction, In *Environmental Effects on Spacecraft Positioning and Trajectories*, pp. 157–73, International Union of Geodesy and Geophysics, 1993.

SamantaRoy, R., Hastings, D. E., and Ahedo, E. A systems analysis of electrodynamic tethers, *Journal of Spacecraft and Rockets*, 29(3):415–24, 1992.

SamantaRoy, R., and Hastings, D. E. A brief overview of electrodynamics tethers, In R. N. DeWitt, D. Duston, and A. K. Hyder (eds.), *The Behaviour of Systems in the Space Environment*, pp. 825–36, Kluwer Academic Publishers, 1993.

Sasaki, S., Kawashima, N., Kuriki, K., Yanagisawa, M., and Obayashi, T. Vehicle charging observed in SEPAC Spacelab-1 experiment, *Journal of Spacecraft and Rockets*, 23(2):194–9, 1986.

Sasaki, S., Kawashima, N., Kuriki, K., Yanagisawa, M., Obayashi, T., Roberts, W. T., Reasoner, D. L., Williamson, P. R., Banks, P. M., Taylor, W. W. L., and Burch, J. L. Neutralization of beam-emitting spacecraft by plasma injection, *Journal of Spacecraft and Rockets*, 24:227–31, 1987.

Shuman, B. M., Vancour, R. P., Smiddy, M., Saflekos, N. A., and Rich, F. J. Field-aligned current, convective electric field, and auroral particle measurements during a magnetic storm, *Journal of Geophysical Research*, 86:5561–75, 1981.

Simons, G. A. Effect of nozzle boundary layers on rocket exhaust plumes, *AIAA Journal*, 10(11):1534–5, 1972.

Singer, S. F. (ed.) *Interaction of Space Vehicles with an Ionized Atmosphere*, Pergammon Press, Oxford, 1965.

Sternglass, E. J. Backscattering of kilovolt electrons from solids, *Physical Review*, 95:345–58, 1954.

Stevens, N. J. Space environmental interactions with biased spacecraft surfaces, In H. B. Garrett and C. P. Pike (eds.), *Space Systems and Their Interactions with Earth's Space Environment*, Vol. 71, pp. 455–76, American Institute of Aeronautics and Astronautics, Washington, DC, 1980.

Stiegman, A. E., and Liang, R. H. Ultraviolet and vacuum–ultraviolet radiation effects on spacecraft thermal control materials, In R. N. DeWitt, D. Duston, and A. K. Hyder (eds.), *The Behaviour of Systems in the Space Environment*, pp. 259–66, Kluwer Academic Publishers, 1993.

Swenson, G. R., Mende, S. B., and Llewellyn, E. J., The effect of temperature on shuttle glow, *Nature*, 323:529, 1986.

Swift, H., Bamford, R., and Chen. R. Designing space vehicle shields for meteroid protection, a new analysis, In *Symposium on Processes for Solid Bodies, Symposium 6, XXIV COSPAR*, COSPAR, 1982.

Szuszczewicz, E. P. Technical issues in the conduct of large space platform experiments in plasma physics and geoplasma sciences, In *Space Technology Plasma Issues in 2001*, Jet Propulsion Laboratory 86-49, California Institute of Technology, Pasadena, CA, 1986.

Tennyson, R. C. Atomic oxygen and its effect on materials, In R. N. DeWitt, D. Duston, and A. K. Hyder (eds.), *The Behaviour of Systems in the Space Environment*, pp. 233–57, Kluwer Academic Publishers, 1993.

Thiemann, H., Schunk, R. W., and Bogus, K. Where do negatively biased solar arrays arc?, *Journal of Spacecraft and Rockets*, 27:563–5, 1990.

Thiemann, I., and Bogus, K. Anomalous current collection and arcing of solar-cell modules in a simulated high-density low-earth-orbit plasma, *ESA Journal*, 10:43–57, 1986.

Tribble, A. C. The space environment and it's impact on spacecraft design, In *AIAA-93-0491*. 31st Aerospace Sciences Meeting in Reno, NV, American Institute of Aeronautics and Astronautics, Washington, DC, January 1993.

Trubnikov, B. A. *Particle Interactions in a Fully Ionized Plasma*, Vol. 1, p. 105, Consultants Bureau, New York, 1965.

United States Air Force, *Space Environment for USAF Space Vehicles*, MIL-STD-1809 ed., Washington, DC, February 1991.

Vampola, A. L. Thick dielectric charging on high-altitude spacecraft, *Journal of Electrostatics*, 20:21–30, 1987.

Van Allen, J. A. *Geomagnetically Trapped Radiation*, John Wiley and Sons, New York, 1971.

Viereck, R. A., Murad, E., Green, B. D., Joshi, P., Pike, C. P., Hieb, R., and Harbaugh, G. Origin of the shuttle glow, *Nature*, 354:48, 1991.

Vincenti, W. G., and Kruger, C. H. *Introduction to Physical Gas Dynamics*, Krieger, Malabar, FL, 1965.

Wang, J. *Electrodynamic Interactions Between Charged Space Systems and the Ionospheric Plasma Environment*, PhD thesis, Massachusetts Institute of Technology, Cambridge, MA, 1991.

Weast, R. C. (ed.). *CRC Handbook of Chemistry and Physics*, CRC Press, Boca Raton, Florida, 1984.

Wei, R., and Wilbur, P. Space charge limited current flow in a spherical double sheath, *Journal of Applied Physics*, 60:2280–4, 1986.

Whipple, E. C. Potentials of surfaces in space, *Reports on Progress in Physics*, 44:1197–1250, 1981.

Whipple, F. L. Meteorites and space travel, *Astronomical Journal*, 1161:131, 1947.

White, O. R. (ed.) The solar output and its variation, Colorado Associated University Press, Boulder, CO, 1977.

Whitten, R. C., and Poppoff, I. G. *Fundamentals of Aeronomy*, John Wiley and Sons, New York, 1971.

Winckler, J. R. The application of artificial electron beams to magnetospheric research, *Review of Geophysical Space Physics*, 18:659–82, 1980.

Wulf, E., and vonZahn, U. The shuttle environment: Effects of thruster firings on gas density and composition in the payload bay, *Journal of Geophysical Research*, 91(A3):3270–8, 1986.

Yeh, H.-C., and Gussenhoven, M. S. The statistical electron environment for Defense Meteorological Satellite Program eclipse charging, *Journal of Geophysical Research*, 92(A7):7705–15, 1987.

Young, D. T., Balsiger, H., and Geiss, J. Correlations of magnetospheric ion composition with geomagnetic and solar activity, *Journal of Geophysical Research*, 87:9077–96, 1982.

Zook, H. A. The velocity distribution and angular directionality of meteoroids that impact on an earth-orbiting satellite, *LPSC*, XVIII:1138–9, 1987.

Index

Made in the USA
Middletown, DE
19 January 2015